BY THEIR OWN HANDS
*A Fieldworker's Account of
The Antigonish Movement*

BY THEIR OWN HANDS
A Fieldworker's Account of The Antigonish Movement

IDA DELANEY

LANCELOT PRESS
HANTSPORT, NOVA SCOTIA

The author wishes to acknowledge the financial support of the Canada Council Explorations Program.

ISBN 0-88999-297-5

Published 1985

All rights reserved. No part of this book may be reproduced in any form without written permission of the publishers except brief quotations embodied in critical articles or reviews.

LANCELOT PRESS LIMITED, Hantsport, N.S.
Office and plant situated on Highway No. 1, ½ mile east of Hantsport.

CONTENTS

Introduction 7
The Big Pay..................................... 13
The Great Default 22
The Study Club — "Ideas with Marrow"............. 33
Meetings.. 42
Something to Read 47
The Short Course................................ 57
A Pocket in a Shirt 66
A New Rochdale or "Turning Merchant" 79
Gentlemen, Let's Have Some Discussion 90
The Big Picture................................. 98
"Women's Work" 105
The Woman with the Basket 117
A Program For Social Change 129
The Keys of the City 134
Waiting for the Snowplow 143
"A Terrible Belief" 150

INTRODUCTION

Students of the Antigonish Movement have access to various sources of information. Its history has been recorded by writers who have dealt with it from many points of view. Its fundamental philosophy can be found in the writings of its founders. Its origins can be traced to the Rural and Industrial Conferences of the twenties. Detailed accounts have been written by the more perceptive of observers who came to the Maritimes with a genuine interest in this unique educational experiment.

Most of these writers, notwithstanding their dedication to the task of faithful reporting, saw the movement from the outside, looking in on the results of the adult education program. They could easily evaluate its worth by making a count of the economic enterprises that were founded as a result of the work done in the study clubs.

This personal account is to describe what went on in the study clubs as seen by this fieldworker. It is meant to be a supplement to the records of the Antigonish movement, describing why people joined the study clubs, what they studied in their small group meetings, how they coped with the problems of the early credit unions and co-operatives, how participation in the program changed their lives.

The decision to write this story was taken in 1982 after a reunion of the women who were on the staff of the St. Francis

Xavier Extension Department in the early years of the movement. In 1982 the annual Atlantic Institute, which was held at St. Francis Xavier University, took as one of the lecture topics: "Women in the Antigonish Movement." Sister Irene Doyle, Kay Desjardins, Zita Cameron, Sister Marie Michael, Ellen Arsenault and I were the lecturers for the week's program. It was a time for reminiscing.

I gave two lectures on my experiences as a field worker for St. F.X. Extension Department in the thirties and forties. Some guests at the Institute suggested that the material in the lectures should be made into a book. Encouraged by the women with whom I worked 50 years ago on the Extension staff, I agreed to write part of what remains in my memory of those early years. We felt a sense of urgency about this effort. Very few of the early field workers are living, and those who remain do not have many years in which to write books. If the story told in these pages is not told now, it will never be told.

I began doing field work, part-time, in Pictou County in 1934 but I had been interested in the movement almost from its inception. The college years 1929-32 coincided with the first active years of St. F.X.'s Extension Department.

The market crash of 1929 came shortly after I enrolled for the college year. The advent of the Great Depression was just in time to become an important subject in economics and sociology classes. During the following three years students were constantly reminded of the economic shambles that prevailed in the country. "Brother, Can You Spare a Dime?" was a popular song. The bread-line was featured in the cartoons of the day and the "queue" entered into daily conversation. The expression "poverty in the midst of plenty" became a cliché.

There were no answers to the vexing questions. Why, for example, should thousands of quarts of milk be dumped into the rivers while babies died of malnutrition? Young people, graduating from university, could make no sense of the way the economic system was run. They could see little progress toward the conquering of poverty.

The era was not without its prophets who proposed remedies for the economic problems. We read about political dictatorships and followed with some dismay the accounts of their rise. In the United States a vast program of public works designed to relieve widespread unemployment was in operation. Keynes' proposals for government intervention in the economic system in Britain was the subject of much discussion pro and con. And eloquent radio orators proclaimed that the economic system would right itself if everybody stopped sinning and prayed more.

The coming of the welfare state was anticipated with enthusiasm by some people and with horror by others. The unemployment insurance plan in Britain, or the "dole" as it was called, caused much controversy. In 1932 it was the subject of the annual intercollegiate debate at Mount St. Bernard College and it was chosen because of the high interest in the topic.

During this time the St. F.X. Extension Department was firmly establishing its program. Hundreds of study clubs were meeting regularly. The year 1932 saw the enactment of credit union legislation in Nova Scotia and the study clubs were ready to start on a decade of rapid organization.

St. F.X. students would have had to be deaf and blind to remain ignorant of what the university was doing in adult education. The Extension program was something new and interesting. It made sense as a means of economic and social reform. Economic cooperation appeared to be a sound basis on which the people of the Maritimes could build the institutions that would provide a better way of life for themselves and their children.

The program was interesting because it was designed to show the people that they could help themselves. It could give each individual participant something practical to do in the work of rebuilding the structure of society. It did not depend on government, either on a democracy inept at curing the depression, or on a dictatorship that could relieve unemployment and poverty only by the destruction of human liberty.

There was something exciting, too, about the way in which the Extension program made it possible for people to work together in a common cause. The possibilities were without limit. Cooperative effort could develop as fast as the intelligence and determination of the people would permit.

The system which had fostered the Great Depression had provided opportunities for rewarding individual initiative and diligence, but the rewards for these admirable qualities were limited to a few favoured persons, and possible only because the masses of the people could not participate in running the business of the country. Cooperation in the economic sphere would permit participation in a democracy not only through political action but also by direct involvement and ownership of a significant segment of the country's economy.

Many of the people who were deeply concerned about the suffering caused by the Great Depression were convinced that the motives of greed and profit had caused the economic breakdown. To replace these motives with those of service and the mutual help inherent in the co-operative movement was a noble idea. It could fire young enthusiasts to work hard for economic and social reform.

During our reunion at the Atlantic Institute we regretted that in the busy years of the movement we had neither the time nor the inclination to keep personal diaries. The St. F.X. Extension archives provide much information but there are many details that have not been writted down for future reference. What is recorded here is from memory but it is the truth. Every occurrence described really took place. I was there when it happened or I heard about it from somebody who was there.

I am grateful for the assistance I have received from my co-workers of fifty years ago. In lively sessions, Sister Irene Doyle, Kay Desjardins, Zita Cameron and Ellen Arsenault reminded me of what life was like in those far-off days in the St. F.X. Extension Department. These pioneer women worked from the opening of the Department's offices as secretaries, organizers, writers and indispensable assistants to Dr. Coady, the director, and A.B. MacDonald, the associate director. Together they possess more knowledge about the Antigonish Movement than any other living persons. They read every word of this book, corrected where necessary, gave advice and their final approval. Without the encouragement of these friends I would never have conquered the inertia of old age long enough to write these pages.

In her lecture at the Atlantic Institute Sister Irene Doyle said of the Antigonish Movement: "It was a wonderful experience. We knew we were involved in something greater than ourselves and that was a source of great satisfaction. We would not have missed it for anything."

In these words she spoke for all of us.

Chapter I
THE BIG PAY

The St. F.X. Extension Department organized its first study clubs at the beginning of the thirties, during the "Great Depression." The people who formed the first study clubs were poor but they had not suddenly become poor with the market crash of 1929. Their poverty was rooted in the economic conditions that had prevailed in preceding decades.

They were poor in the rural communities where the families of farmers and fishermen depended on the products of land and sea for a livelihood. These farmers and fishermen had never had any control over the prices they received for the products of their labour and their investment in farming and fishing equipment. As Dr. Coady put it they had not been able to "say their say" in the transactions of selling and buying. They took what the buyer offered when it was time to sell and they paid what the seller demanded when they went to buy. When the fish buyer was also the community merchant the fishermen often found that there was very little cash left when the books were balanced at the end of the fishing season.

Farmers had no effective organization to pool their resources for the marketing of farm produce. The system of individual marketing resulted in what Dr. Coady called "driblet" farming that was inefficient, and inadequate to provide a comfortable living on the farm.

In 1934 two dollars a day was considered a good wage in rural Nova Scotia. Highway workers got twenty-five cents an hour for a ten-hour day on a highly prized government job. The pay of woodsmen varried. A lumber operator could get all the workers he needed just for their board and lodging. The skilled workers, as sawyers in a mill, received about seventy-five cents a day in addition to their keep.

Farmers could produce enough food for their families but some cash was needed for clothing, taxes, and such luxuries as tea, sugar and baking powder. In a year of a bountiful harvest in wild strawberries and raspberries farm women regretted that they could not preserve the delectable fruit for winter use because they could not afford the sugar.

During the worst period of the depression a rural physician said that he had patients who could not afford to mail a letter. Postage was three cents, first class.

The prevailing poverty drove young men and women from the rural communities to seek employment, the boys in factories and mines, the girls in the homes of affluent American employers. Many a farm family depended on these young women for cash contributions.

Towards the end of the twenties there began a movement to bring the plight of the fishing communities to the attention of the federal government. The first public stirring to this end took place in Canso in 1927. Dr. J.J. Tompkins, one of the founders of the Antigonish Movement, was the instigator of a public meeting held in Canso on July 1, Dominion Day. It was the year of the Diamond Jubilee of Confederation and the event was marked by celebrations throughout the land. The people of Canso did not join in the festivities. Their meeting was not for the purpose of waving the flag and listening to patriotic speeches, but to discuss what could be done to help the poverty-stricken community whose residents found no cause for rejoicing in the Diamond Jubilee of Canada.

From this meeting a telegram was dispatched to Ottawa, stating that, on this Dominion Day, there was no joy in Canso, and asking what the government intended to do about the desperate conditions in the fishing communities. The Canso meeting received wide publicity. The protest was well timed and forced the government to take notice of the plight of Maritime fishermen. Eventually, the government appointed a Royal Commission to investigate the fishing industry. One of the recommendations of the Commission was that the Maritime fishermen should be organized and one year after the Canso meeting Dr. Coady was appointed to carry out this organization. It formed the basis for the St. F.X. Extension Department's promotion of cooperative enterprise in the Maritimes.

If the people in the rural districts were poor, their cousins in the towns were poorer. At the worst of times the farm family could produce food for the table but the industrial worker's family was entirely dependent on the daily wage for all the necessities of life.

In the industrial districts of Cape Breton and Pictou Counties, where the early study clubs were formed, the basis of the economy was in coal and steel. These areas had been hard-hit by the depression. Loss of markets had brought severe unemployment and "slack" time. In the coal and steel areas the depression made life harder than ever but the miners and steelworkers, like farmers and fishermen, had for a long time suffered from a heritage of economic troubles.

The decade of the twenties had been one of great hardship for the miners and steelworkers who had struggled against wage reductions, and employer opposition to their unions. In the coal towns the year 1923 saw the beginning of a series of wage reductions which the miners tried to resist by means of strikes. Each time they were forced back to work without gaining anything.

In our time the public has become accustomed to strikes that are called by unions that demand higher wages, shorter hours, better pensions and other benefits. The strikes by miners and steelworkers in the twenties and thirties were desperate efforts to avoid wage reductions. Today, unions are engaged in making life easier for their members. Unions in the industrial areas of N.S. in the twenties and thirties were trying to keep life from becoming *worse.*

The coal miners' strike of 1925 had a devastating effect on the coal towns. The miners stayed on strike for five months and were finally forced back to work with a ten per cent reduction in wages. The dispute between miners and their employers was settled not by negotiations at the bargaining table, but by a provincial army of soldiers on foot with gun and bayonet, and soldiers on horses that rode ruthlessly through assembled workers. The memory of the horrors of 1925 remained forever with the families that lived through the strike. That memory was still vivid in the minds of the many miners who visited the Extension office in Glace Bay in the mid-thirties. From these men I learned what life had been like in the time of the strikes.

The only glimmer of hope left to the miners who lost the strike of 1925 was the promise from the provincial government that it would appoint a Royal Commission to investigate the coal industry. The Commission's terms of reference were broad. It was ordered to inquire into the wages and conditions of labour, social and domestic conditions, the causes of strife between employer and employee, the general financial arrangement and cost of management, market conditions and any other matter deemed by the Commission as affecting the industry.

The eagerly awaited report of the Commission was a sad commentary on life in the coal towns. The Commission found that low wages, insecurity of employment, "deplorable" housing, the opposition of management to the miners' union and radical agitation within the union all contributed to a depressed standard of living in the mining communities. Yet in spite of this declaration of misery, the Commission said that the ten per cent reduction which the miners had tried to resist in 1925 was "amply justified"!

The remainder of the decade brought no improvement in the harsh life of the miners. In 1932 a second Royal Commission was sent to investigate, at a critical time when the miners were being threatened with another wage reduction. The report of this second Commission was a gloomy document. It grudgingly admitted that the miners were "probably" right in expecting that when the cost of living dropped, they should be able to retain their wage rate, thus making a little progress toward raising their standard of living. But the Commission recommended a wage reduction only a few cents short of that proposed by the company, the closing of five mines which employed a total of 2500 men, and part-time for those who were left on the employment rolls.

The depressed conditions continued into the rest of the thirties. There was a slight recovery in 1937 when new markets for coal were found, but in 1938 coal suffered a setback. According to a report by the director of the Coady Credit Union at the 1938 annual meeting, there were weeks when the miners did not get one shift.

Mines designated by the Royal Commission were closed and there was no alternate employment for the displaced men. The mines that remained in operation were on part-time and there was no improvement in wages. The term "part-time" used by the Commission did not indicate the severity of the depression. In real life part-time in the mines consisted of one day a week, two days in a good week and at worst only two days in a whole month.

From the weekly pay there was a long list of deductions for various purposes, including prepaid fees to doctors, hospitals, church, relief society, union dues and "poll" taxes, and most of these had first call on weekly pay. The first Royal Commission had expressed concern over the number of payroll deductions. It found that the average check-off took from one quarter to one third of the weekly earnings, and that, when the miner worked part-time, the total check-off often reached fifty per cent and sometimes one hundred per cent of the miner's pay.

The story has often been told of the miner who on one occasion did not bother to call at the pay office because, from past experience, he knew that there would be no pay to collect, the earnings having been eaten up by the check-off. He erred. There *was* a pay and he was called in by the clerk to withdraw it. After the deductions he was entitled to two cents, which had been officially recorded and enclosed in the envelope. The incident earned for him and his descendents the nickname of "Big Pay."

In those part-time days the miner's family anxiously awaited the sound of the colliery whistle. It blew at six o'clock each evening — one blast to announce there would be work, two that there would be no work on the morrow. The miner's family never knew in advance on which days of the week the mine would operate. This insecurity caused many problems.

There was, for instance, the young people that had fixed the wedding day. The bride's father and brother were both miners and the family hoped that the week's one working day would not coincide with the wedding. On the eve of the wedding the whistle blew for work. The father lost the shift to give his daughter away, but the brother who had a young family to support was not able to attend.

There was no way to replace income lost in time of unemployment. The miners had no savings because there had never been a time in the turbulent twenties when the miner's family had any surplus from which to save. The Royal Commission of 1932 noted this fact, citing the extreme hardships of the twenties during which there had been two wage reductions.

In some communities the situation became so serious that the Nova Scotia government found it necessary to provide some form of assistance. This was known as "relief" and it was administered by municipal authorities in a haphazard manner. A survivor of those hard times recalls that he received one dollar in relief in a week when he had not worked one shift. He had a family of small children to support.

There were soup kitchens in the coal towns. School children were fed raw cabbage and buttermilk during the recess period in an effort to supplement their meagre diets.

The Royal Commission of 1925 had suggested that the miners should make more use of the land. When the men talked about this in their study clubs years later they noted that it was easy for college professors and industrialists to offer this advice. Not much land was available to the miner who lived in a row house with a backyard just large enough to serve as a site for the ash-heaps from the coal stoves. They said that the commissioners did not understand the vicious circle of poverty. The income from one day's work per week did not leave much with which to pay for fertilizers, seeds and garden equipment. The "Big Pay" was proof of that.

Nevertheless, by 1932 there was more use of land. Where a common pasture could be found for the summer, some miners managed to keep a cow, finding a place for the precious creature in the backyard barn in winter. The milk was shared with cowless neighbours. I have heard miners say that the milk, in an unpasteurized form, saved the lives of many babies. When laws requiring pasteurization were passed the miners' cows disappeared.

The Nova Scotia government tried a few experiments in placing unemployed miners and steelworkers on vacant farms. The plan was doomed to failure. These farms were vacant because farmers had not been able to make a living on them. It was too much to expect that a coal miner could take a rundown farm and make it pay.

There were some interesting stories about this experiment. A favourite was that of the miner who, after a complete failure on the farm project, was left with nothing but a horse, so he rode back to his mining town on horseback. He was penniless but he had transportation to take him back to the place from which he had come.

Steelworkers who started early study clubs had a history much like that of the miners. They too had lived through strikes, wage reductions and the presence of the army in their streets.

There is a difference between the conditions of the poor in town and country in the thirties and the poor of the eighties. By our standards their wants were simple. There were no payments to be made on cars, stereos and electrical appliances because the poor people had none of these things. They had no phone bill because in the community only the doctor, the clergyman and the merchants had telephones. The wants of the farmers, fishermen and industrial workers during the hard times of the depression were not very great — a warm home, sufficient food and clothing, medical care. To lack these things was to live in poverty.

There were no retirement pensions, no family allowances, no medicare, no unemployment insurance, no social assistance as we know it today. Old age pension legislation was introduced in 1934 with a maximum payment of twenty dollars a month, with a means test, which made the pension available to paupers only. One old lady who owned a piece of land which produced nothing by way of income except taxes to the municipality, was allowed a pension of twelve dollars and ninety cents a month. The reduction from the maximum twenty dollars was made because she was classified as a "property owner."

Unemployment insurance was a dream, a system advocated by people who were considered to be irresponsible radicals. Arguments for and against the concept of unemployment insurance were frequent, based chiefly on an evaluation of the English system or the "dole" as it was called. Opponents of unemployment insurance were sure that nobody would work if an income could be obtained during idleness. Advocates of unemployment insurance compared this stand to the arguments of the opponents of workmen's compensation, who claimed that workers would cut off their hands and feet to get benefits.

The blind, the aged, the disabled were cared for by their families, often with the assistance of neighbours. It was a time of mutual help when neighbours shared with others what little they had.

Into these impoverished rural villages and industrial towns came Dr. Coady with a message of hope and a practical plan of adult education and economic co-operation. No wonder that the study clubs were adopted with enthusiasm!

This was especially true in the towns. When men were employed only one or two days a week there was plenty of time to spend in study club meetings. The topics to be discussed in the clubs were the economic problems that beset the workers' families. The rising bank of coal for which no sale could be found, the constant struggle to pay for the week's groceries, the problems in the labour unions,— all these were the topics of conversation where people met on the street corner, in the community hall, around the stove in the kitchen.

The rural people had their own special economic problems. With their city cousins they shared the constant worry of managing to feed, clothe and shelter a family, to provide it with needed medical care and to make provision for children's education.

In the clubs there was now a chance to study these matters in an organized way, under supervision, and with an abundance of reading material.

When Dr. Coady talked to the people about "mobilizing for action," he often stressed what he called the "force of ideas." He said that the program of adult education *had* to result in the rise of institutions that would eventually enable the people to conquer poverty and attain the "good life." Nothing could stop the force of ideas. Dr. Tompkins said the same thing another way. "Ideas," he said, "have hands and feet. They work for you."

There is a striking example of the force of ideas at work in the industrial workers' study clubs. The first credit union was organized in 1933 and it was followed rapidly by others throughout the industrial areas. During the previous year, 1932, the clubs were busily making a study of the credit union, preparatory to the process of organization.

But 1932 was a critical year, the year of the second Royal Commission on coal. Learned men accepted as inevitable the deplorable standard of living in the mining towns, and recommended the closing of mines with a consequent frightful increase in unemployment. But while they wrote their document of despair, the victims of whom they wrote — men unemployed or existing on two work-days a week — sat in their study clubs and planned how to start cooperative banks. It would be difficult to find in the history of Nova Scotia a more striking example of courage in the face of economic adversity.

Chapter II
THE GREAT DEFAULT

In instructions which he prepared as a guide for speakers, Dr. Coady summarized the philosophy and practical objectives of the St. F.X. Extension movement. It was a program of adult education by which the people of the Maritimes would learn to use their resources to conquer proverty and bring about the social and economic reforms that would give them the "good and abundant life." Its philosophy was one of love and brotherhood.

The application of the program in the Maritimes had to be positive and concrete. There should be no preaching of social reform, in abstract principles, as from a lofty tower. What was learned in the study clubs would be applied to the building of economic institutions owned and controlled by the people who had in themselves the power to make the Maritimes great. From the beginning there would be a sense of achievement.

The people would learn the techniques of group action and apply them to the building of their own co-operative institutions, in credit unions, in marketing and consumer co-operatives and would progress from retailing to processing, transportation and manufacturing of consumer goods.

The program was intended to give everybody the opportunity to participate fully in the economic life of the community and province. In order to succeed the program had to develop leaders prepared to fill responsible positions in the new economy.

Dr. Coady looked upon the Maritimes as a social laboratory, large enough to be meaningful, small enough to be manageable, an economic unit that had all the ingredients to become great. Even in areas not well endowed by nature, where the soil was poor and the winds blew cold there were wonderful human resources — men and women using their intelligence and hard work to achieve a good living.

This was the program which Dr. Coady, at public meetings, introduced to audiences that never failed to be stirred to action by his analysis of their problems, the logic of his presentation and the eloquence of his words. Dr. Coady had a high regard for the general or "mass" meeting, as he preferred to call it. He would say that the mass meeting was the place to explode the intellectual dynamite that would break up apathy and prejudice, shock the people out of their complacency and fire them with enthusiasm to rebuild society. He liked the analogy used by Bishop Grundvig, the great Danish educator, who compared the meeting to an alarm clock. "The clock is useful to wake you up in the morning," said Dr. Coady, "but you don't want the alarm to ring all day. You have to get up and go to work."

Many of Dr. Coady's speeches, those delivered at conferences throughout the continent, have been recorded and preserved. They were carefully prepared and written in advance of the occasion on which they were given. But what he said to the farmers, fishermen and industrial workers, in schoolhouses, community halls and local theatres, has been lost forever. The few notes, scribbled on a torn piece of paper, were discarded after the meeting. There were no tape recorders in the early years to preserve those speeches for posterity. All that remains of the meetings where he taught his philosophy directly to the people, is in the memory of those who heard him and of the workers who accompanied him to take part in the program.

I was fortunate to be one of those workers, to observe how Dr. Coady inspired and challenged his hearers to unite in a program of adult education that would change their lives.

How he did this can be described by recalling a series of meetings held in the rural areas of Antigonish and Guysborough Counties on each of six successive evenings. We were accompanied on this tour by some students from the St. F.X. sociology classes and, on several evenings, by visitors from the United States.

At these meetings, Dr. Coady introduced his theory of the "Great Default," which he later included in his *Masters of Their Own Destiny*. During the ride on the first evening he began the explanation of this theory, but there was not enough time to develop it fully, not with his new Buick purring along at maximum speed. There was just enough of the "Great Default" to make us anticipate a rousing speech.

The meeting having been duly opened by the chairman, Dr. Coady would announce to the people in the assembly that he was about to put them in a state of scientific humility. The astonished farmers would sit up straight at the school desks, somewhat puzzled as to what to expect. The scientist, Dr. Coady would continue, in his search for the truth is humble, awed by the mysteries of the unknown and the insignificance of his own achievements in the vast world of knowledge. The more he learns, the more humble he becomes and the more eager to learn even more. Dr. Coady expected the people in the meeting to be humble enough to acknowledge how little they knew about the economic forces that governed their lives, — the financial system, the workings of the marketplace, and their own lack of knowledge of scientific agriculture. When they became ready to admit how much they needed to learn to be successful in their avocation they would be ready to open their minds to new ideas. They would be in a state of scientific humility.

Sometimes Dr. Coady would express a desire for a magic iron with which to iron out the crooked thinking from their minds as they filed past him one by one. At other times he would switch appliances and suggest that a giant vacuum cleaner would be great to gather up all the stupid ideas abroad in the place. Since no such mechanical device was available, he would propose a slower method, one sure to work. It was adult education.

This kind of education began on the spot as Dr. Coady proceeded to explain to the farmers what was wrong with the system that made them poor. He was about to explain the "Great Default" and the university students and visitors took out paper and pencil. I observed that they took few notes on the first evening. They just sat there and listened, spellbound, as Dr. Coady unveiled his "formula" in a speech livened by illustrations which he called "parables."

He would start with a vivid description of the hard life of the pioneers who built their log homes with their own hands and toiled from dawn to dark to wrest a bare living from the land. In the beginning everybody shared in this daily work, but there was one lazy man who particularly disliked all this effort. Besides being lazy he was also smart, thus possessing a great combination of the qualities required by a rugged individualist. After a period of deep thinking, the lazy man hit upon a good way to avoid all that hard work and, at the same time, live better than everybody else. His bright idea was to supply his neighbours with the things they needed and to be paid for his services. And so the system of merchandising began in a one-room shack equipped with a few shelves and stocked with staples that were needed by each family. Behold! The merchant appeared on the scene.

The people of the community would then come to this smart fellow's store to exchange their products, or their money when they had it, for the goods which they were not able to produce for themselves. But there was something inherently wrong with this transaction. It was a unilateral, not a bilateral agreement; therefore it was not a contract. The buyer paid the price demanded by the seller. "Right there," said Dr. Coady, "our forefathers let go their rights as consumers to have some control in the transaction between buyer and seller."

The same thing happened in the field of finance. Banks began in the same humble way and the people let go their right to control their own money.

This was the great default of the people. *They failed to get control of consumer business and they failed to get control of finance.* The primary producers who create the wealth of the country lost their right to "say their say" in the control of economic institutions and, because they allowed this to happen, the economic system was built on the wrong foundation.

Then came the "leaning tower" parable. "The economic system," said Dr. Coady, "is built like a tower that is out of plumb." Everybody in the audience understood that. The farmers had all, at some time, built barns or sheds or chicken coops, and they knew how important it was to start the walls rising up straight, not listing to right or left. Perhaps some inept farmer had once raised a structure that was out of plumb, thereby causing much merriment on the part of all who passed by.

"When a leaning tower rises high enough," said Dr. Coady, "it becomes menacing and attempts are made to prop it up with guy-wires. But no matter how many wires are used, the tower still leans. If the building process continues and the tower is allowed to rise high enough, the time will come when nothing will keep it from crashing down. Society is making feeble attempts to prop up the leaning economic tower with the guy-wires of a skimpy old age pension, mother's allowances and the rusty old wire the dole. But the fact remains that the economic tower still leans, no matter how many attempts are made to hold it up. The only wise course is to rebuild the tower on a true foundation. Only in this way will the people correct their great default."

Sometimes Dr. Coady would find this a good place to talk about the radical views of those who professed to fix that tower fast. This was the revolutionary way, the way of the communist who wants the state to take over the system and destroy the tower in one grand blow-up. The communist, says Dr. Coady, is like the city dweller who moves to the rural community with plans to make farming pay. "He'll show you boys how to speed things up," explained Dr. Coady, "and so he puts two hens on the nest. But you and I know that the chickens will not hatch any faster with two hens sitting on them than with one." The whole meeting would explode with laughter. Two hens on the nest? I used to think that those farmers would ever after conjure up the picture of a two-hen nest when they heard the word "communist."

After this entertaining digression into communism, Dr. Coady was back at his leaning tower. The people of the Maritimes would not obtain the good life by any more attempts to prop up the menacing structure. The scientific way was to rebuild the system on a solid foundation and see that it rose straight and true from its base.

It was at this point that Dr. Coady would speak of the resources which the people had at their command. He would give a list of the good things which rural dwellers had a right to enjoy: economic security acquired from farms that produced in abundance and marketing agencies that assured a fair return for the farmer's work; good homes with electricity and other modern conveniences; healthy families that are well fed according to the latest scientific knowledge. What fun he had describing the well laden farm table! He never failed to mention strawberries and cream. Every farm family should have lots of big juicy strawberries, fresh from the patch in summer and from the freezer in winter. Those critics who said sometimes that Dr. Coady had his head in the clouds never heard him tell his audiences how good life could be on the farm.

And then, because the Nova Scotia farms were so far from reaching the standard he had set forth, Dr. Coady would administer a scolding, "laying on the rawhide lash," as he called it. He would describe how inefficient the farmers were and they would laugh at his description of a farmer, with nothing but a pig to sell, complaining of the lack of markets.

Sometimes he would make his criticism indirect with his parable of the Patagonians, a trio of fictitious visitors from a far-away land who accidentally go off course and arrive in their ship at the Maritimes. They go ashore and visit for a short time, but they withhold their comments until they sail away for home. When they talk among themselves they shake their heads at the stupidity of the inhabitants and their strange ways. They are surrounded by magnificent forests but they live in poor homes, some of them little better than shacks. They are scrawny and lantern-jawed from poor nutrition but they have lots of land on which to grow an abundance of food. Some are hungry because they are unemployed but there seems to be plenty of work to do. Some of these strange people build their houses on the top of a hill, dig a well at the bottom and for the rest of their lives haul water in buckets up the hill. Evidently they are not smart enough to reverse the plan, built the house down below, dig the well at the top and pipe the water down.

The Patagonians sail away, feeling sorry for the stupid Maritimers.

Since by this time the farmers were in an acceptable state of scientific humility, Dr. Coady would do what he called "pouring oil in the wounds" caused by the application of the rawhide lash. He would present his "formula" by which the people of the Maritimes could begin to recapture the rights which they lost by their great default. The basis of the formula was adult education. He would remind his listeners that the world is run not by children but by adults, and it is therefore foolish to live as though education ends for each person on the last day of school. One of Dr. Coady's frequent statements was that education should be co-terminus with life.

He would explain how the educational system failed to prepare farm families for their avocation. He would describe how a family made sacrifices to send the brightest boy to university to prepare for one of the professions. (This privilege was never extended to a girl. The girls were in the United States, sending money home to help pay the college expenses.) How the members of the family would glow with pride when news came that the favored son had received his degree at exercises which they could not afford to attend! But this education was a door through which the favored few escaped. The sad thing was that it was a trap door opening only one way, and the boy, educated at so great a cost to his family, never came back. He was lost to the community forever.

People in rural areas should no longer be content with trap door education, Dr. Coady said. Adult education, which he proposed, was for everybody and would be carried on through study clubs where the people would learn how to help themselves and others to rise to a better life. Their study clubs would lead them in the pursuit of certain specific goals.

The first goal was the attainment of personal individual efficiency for every farmer. "Every human being," said Dr. Coady, "should be efficient enough to carry on his vocational activities in a scientific rational way." For farmers this meant the acquiring of a sound basis in agricultural methods that would make the farm as productive as possible and rural life good and happy.

But if the first goal was self-help, the second, equally important, was mutual help among the rural people. Group action was absolutely necessary. No matter how efficient the farmer became, the marketing of his products could be done effectively only in joint effort with the other farmers in the community. Individual efficiency would bring the most returns when it was accompanied by group efficiency as farmers banded together to carry out those economic activities which one farmer could not do alone. The possibilities for group action were many and included all phases of marketing, processing and transportation of products.

But the farmer who would stop at cooperative marketing was unscientific. "Cooperative marketing," said Dr. Coady, "will bring the farmer only as far as the vestibule of cooperation." It was consumer cooperation that would correct the great default in merchandising. "The co-operative dream," said Dr. Coady, "is to cover the Maritimes with cooperatives and set the wheels of industry turning in our own factories. Then the consumer will be in the driver's seat." He foretold that the scope for group action in co-operatives was unlimited and could be extended by a resourceful people into services including medicine, housing, libraries and recreation. At this point he would tell the story of the Rochdale Pioneers and urge the farmers to follow their example.

He would define clearly the special function of co-operatives in the Antigonish plan. There was some misunderstanding about this among the non-believers and critics. Although Dr. Coady wanted to see the Maritimes covered with co-operatives, he never advocated nor did he forsee a total co-operative economy. Such a result was neither possible nor necessary. There could be co-operatives all over the Maritimes and plenty of other business too.

In the Extension plan co-operatives were to form an important component of the total business of the country. The goal was to obtain a percentage of business sufficient to give co-operators the power to control prices. When co-operators could demonstrate their ability to make a success of manufacturing, they could be a powerful check on monopolies. Sweden was the prime example of a country where this had been accomplished by co-operators. The Swedes were renowned as trust busters. Dr. Coady believed that Maritimers could be trust busters as well as the Swedes.

The remaining business would have ample opportunity to grow in a society where the standard of living was rising as a result of cooperative effort. It might still comprise the larger portion of the total business but it would be good business, held in check from rampant profit-taking by a vigorous co-operative sector.

Although Dr. Coady had succeeded in placing his hearers in a state of scientific humility, he did not intend that the state should be permanent. It was time for them to come out of their humble state and aspire to do great things. He believed that they could succeed and they knew he believed it.

We heard the "Great Default" speech for the next five evenings but did not grow tired of it. It grew better and better as the week passed. Some evenings I enjoyed more than others. I had a part in the program. This was a talk on details of the Rochdale principles and the place of consumer education in the co-operative retail stores.

Sometimes Dr. Coady would elect to speak last on the program and that was how I liked it because I could say my piece and then relax and enjoy the rest of the meeting. But sometimes he would speak first. It was a worry to have to speak after the audience had been treated to intellectual dynamite exploding all over the schoolhouse.

But Dr. Coady was the kindest of critics. It was always a source of wonder to me that so gifted an orator had never a critical word to say about a speaker no matter how poor the performance. He would sit there, beaming with pleasure, when somebody from the audience would speak a few halting sentences in support of the movement.

The tour of the rural areas was followed by a similar tour in Cape Breton industrial areas. The town meetings heard the "Great Default" speech with the parable of the leaning tower. But when Dr. Coady spoke to wage earners he emphasized the importance of consumer co-operation, as a way to give them economic power.

The rural people were owners of property. They owned their land, their homes, their equipment and their stock. If they owed debts on these things, they still had some equity and they surely owned the products they grew on the land. In contrast, the wage earners owned little or nothing. With few exceptions, they lived in rented homes. They did not own the products of their labour.

Dr. Coady very emphatically told them that they deserved more than the wages of their daily toil. They had a right to some degree of ownership and participation in the economic system under which they lived.

The way to obtain this ownership was through consumer co-operation or "labour's other arm." The only thing the worker owned was his wage, and over that he had complete control. When the pay went into his pocket he could decide to use it in a way that gave him ownership in the credit union through his savings and in the co-operative store through spending.

The message was clear and concise. "Begin where you are," Dr. Coady said. "Build your credit unions and your co-operative stores. *Strike your own little blow.*" The "little blow" became a slogan for co-operative workers. Dr. Coady never failed to arouse the enthusiasm of his audience when he predicted that together the little blows struck by co-operators all over the Maritimes would result in the "sledge hammer lick" to change the world.

The acquiring of ownership was one of the most powerful themes of Dr. Coady's speeches and the idea was one of the most frequently discussed in the study clubs. He believed that pride of ownership ws essential to the dignity of the person. "There is no democracy without ownership" he repeated from many platforms.

He was at his most eloquent when he told the wage earners that ownership in the economic processes of the country would bring them personal rewards as well as monetary gain. His imagery was striking: The miner emerging from the black pit into the light of day, straightening his aching back, looking up at the sun and saying: "I am a banker; I am a merchant; I matter in the economic scheme of things."

Whenever he spoke at meetings in town or country Dr. Coady communicated to his audiences the firm belief that participation in the program of adult education would change not only the economic order but the people themselves. "The doing of it" would be as important as the final result. He regretted the waste of talent in a society where the people had not been free to reach their potential in learning and group action. "The saddest thing that can be said of a man," he would say, "is that he was born, he lived and he died, with no accomplishment toward a better social order between his birth and death." Over and over again he said "Co-operation builds men," and he could prove it.

It is almost fifty years since that particular speaking tour of Dr. Coady's took place. It is an indication of how profoundly impressed I was by the experience that I could reproduce the variations of one of his speeches from start to finish without any reference to a written record.

Chapter III
THE STUDY CLUB
"Ideas with Marrow"

The study club was the fundamental unit in the Antigonish movement. Everything that was accomplished in the experiment began with the free discussion in the weekly gathering of friends and neighbours.

The study club method was selected because it was practical. There were two requirements for the success of the program. The first was that it should enlist the participation of large numbers of people. This was to be education for the many, not for the few. The second was that there should be continuous regular periods of study.

The study club was practical because it did not require full-time teachers and there was no need for students to move from their communities for instruction. The students would teach themselves. With its small staff the Extension Department could carry out the initial organization and provide study material and adequate supervision.

The founders of the movement were confident that they would succeed if they held fast to the principle that the clubs had to pursue some practical concrete goal. As Dr. Coady put it: "Study must issue into action." The men and women in the study clubs began with the specific goal of building co-operative enterprises. To enrol hundreds of farmers, fishermen, miners and steelworkers in this program was what Dr. Coady called "mobilizing for action."

The way in which this was done was simple, direct and swift. If a program of this kind were undertaken today, it would probably require a long period of preparation. First there would be a feasibility study. A task force would examine all aspects of the proposal. Psychologists would be called in to report on the learning abilities of the prospective students. Consultants and statisticians, equipped with their computers, would have much work to do before the first student received a leaflet to study. It is well that the Extension program started without much preliminary investigation. Otherwise a long time would have elapsed before the opening of the first credit union.

The founders of the Extension movement went directly to the task. They assumed that the people would rise to the challenge. Dr. Coady was confident that they could do great things if they were shown the way.

It is true that the social conditions of the time made the study club attractive. It would be difficult today to find large numbers of people willing to give up an evening a week to a study club. Everybody has too many meetings and social activities to attend. In the thirties there were few distractions to keep members away from the study club meeting. There was no television and there were few radios. People had nowhere to go, no cars in which to drive anywhere, and no money to spend. The study club meeting was often the social highlight of the week. The clubs usually met in rotation at the homes of the members and the study period was combined with a friendly visit, but there were variations from this procedure. Clubs met in odd places.

There was, for instance, a lively group of young women who decided that they needed a special meeting place. The father of one of the women owned one of the few cars in the community and this car had a garage of its own. The girls scrubbed it clean, put curtains at the windows and pictures and posters on the walls. They collected odd bits of furniture from their attics, repaired and painted them and found used mats for the floor. On study club night they pushed the car out, laid the rugs, set up tables and chairs and presto, an instant club house.

There was one unusual meeting place on record. An enthusiastic study club member got into difficulties with the law. He was the community's most upright citizen but in a momentary fall from grace he shot a deer out of season. On a matter of principle he refused to pay the fine and landed in the county jail. A.B. MacDonald was in the area for a meeting and he and his travelling companions decided to visit the prisoner. a very good friend of theirs, with the purpose of cheering him up in his lonely cell. They were surprised on arrival to find that he did not need any cheering up at all. He was delighted to see AB and his first greeting was: "Send me some pamphlets. This is a great place for a study club." He was not about to spend his sentence in idleness and had organized the prisoners.

Many of the study clubs originated at general meetings where, after speeches designed to persuade the audience to join the educational movement, leaders were chosen. Usually these leaders were volunteers, but if volunteers were slow in speaking out, the meeting would choose them. It was always interesting to observe how unerringly the meeting would choose the best leaders. Sometimes we cheated a little in describing what was expected of a leader, by making the position appear to be easier than it actually was. Prospective leaders were told that their role was simply to be a link between the study club and the Extension Department, to receive and distribute literature and to report on the progress of the club. In practice the leader was expected to take charge of the meeting and to lead in the discussion.

The study club was made up of neighbours and friends and so represented a cross section of the community. In each club there were those who spoke too long and too often and those who spoke rarely or not at all, although experience tended to loosen the tongues of the shy ones. There were the rash members who were daring and eager to embark quickly on all kinds of projects and those who were over cautious about engaging in group action. There were pessimists and optimists, and between the two extremes were the members who kept the club on a reasonable course toward progress.

The Extension Department produced and distributed a pamphlet giving directions on conducting a study club and leaders were expected to give careful attention to this little book. The directions in the study club pamphlet were to be used as guidelines. No hard and fast rules were set, but it proved to be of much help to the clubs and it was widely distributed throughout the Maritime provinces.

Most of the study clubs began their first meeting with a leaflet entitled: "How We Came to Be As We Are" written by George Boyle. This was a sure-fire topic. Those members who heard Dr. Coady explain his "Great Default" theory already had some ideas of the causes of their economic troubles. George Boyle's "How We Came Hither" was another first meeting leaflet.

The first questions asked in the study club were: Why are we so poor? Why can't we make the farm pay? Why the spread between what we get for our fish or our vegetables and the price which the people in the town pay for these products? Why are we always in debt?

The study clubs in the towns asked their questions in a slightly different way: Why can't we afford to buy the food we need for our families while the farmers have to sell their products at such low prices? Who gets the profit in between the two transactions? Why are we always in debt?

These questions led to others prefaced by "How?" instead of "Why?". How can we get out of debt? How can we get better prices? How can we make our wages buy more? How have others helped themselves?

In the rural clubs efficiency in production, attention to scientific farming and group action in marketing were subjects for study as soon as the groups became organized. The support of the government agricultural representatives was an important contribution to the success of the study clubs.

In the towns the clubs did not have as much opportunity for group action as did the farmers and fishermen. The life of the industrial worker was dominated by the established corporations. Group action as producers was limited to labour unionism. Immediate group action as consumers was the first topic for the industrial clubs, and the history of the credit union movement and of economic co-operation became the chief topic of study.

The questions referring to local conditions led the study clubs to discuss the structure of the economic system. In the towns the clubs were drawn into a more intensive study of contemporary social movements and the relative merits of diverse economic systems.

Much of the planning for the credit unions was done in the meetings of the Associated Study Clubs but the preliminary study was done in the individual clubs. It was in the clubs that the members learned their lessons on credit union and co-operative legislation, on the structure of credit unions and co-operatives and on the requirements for the proper management of these enterprises.

The easy way to evaluate the results of the St. F.X. Extension program was to count the number of credit unions, co-operative stores and fish plants and to add up their assets in dollars and cents. These were the visible results. But there were other results not so easily calculated but equally and perhaps of greater importance. There was the personal enrichment brought into the lives of those who participated in the program. The study clubs gave men and women a way to continue the education that had been abruptly cut off when they left school.

There was a myth about young people coming out of school from early grades with whoops of joy. Unfortunately there were too many boys and girls, talented and eager to continue their education, who envied their more favoured companions who remained in the classroom with prospects of reaching graduation day.

There were very few high school graduates in the study clubs. Economic necessity had forced many of them out of school. Young people had to help, where possible, in earning the family income. Girls were kept out of school to help look after younger brothers and sisters or to work as hired help for the small wage which that job provided. Young men left school to look for work in mines or steel plants or anywhere it could be found.

Michael McNeil, one of Nova Scotia's most prominent credit unionists, had gone into the mine at the age of thirteen to help support a large family of brothers and sisters. His father, after a mine injury, was unable to return to this job. In later years, Michael McNeil would still relate that he had left school with real regret. He would add that when the study clubs were organized in New Aberdeen his new life outside of the mine began.

Michael McNeil was a charter member of Coady Credit Union. For fifty years he gave that credit union devoted service. He was its first secretary who rarely missed a directors' meeting and in whose hand the records of the credit union for two decades are written. He served as clerk from the very first day, as the unpaid assistant manager and finally as manager of one of the province's most successful credit unions. The two names by which he was affectionately known in the community sum up the life of service of an extraordinary co-operator. He was UMW Mick and Credit Union Mick and justly entitled to both names.

There were few opportunities to continue education for those who had to drop out of school. In the thirties night schools were limited to courses in trades and these often required basic education which applicants did not possess. Thus, the study clubs became an exciting experience for those who had lost out in the educational system. For the first time since leaving they had something to read. A new world was opened to them when pamphlets and books became available and many study club members formed reading habits that lasted all through life.

The study club members enjoyed the stimulation of group discussion. What education they had received in school had not been through discussion. Memorizing and reciting lessons had been the techniques by which they had passed from grade to grade and out of school forever. The study club was a little forum where, in free discussion, the members gained the confidence that enabled them to speak in public when the need arose. At a conference at Antigonish in 1942 Sister Marie Michael said: "There are men and women in this room who ten years ago had not made a public utterance in their lives, but who now can hold their own in any public meeting." Heads nodded in agreement.

The study club was particularly interesting to its members because it did not function in isolation, but in common with hundreds of other clubs, all with the same objective, as part of an important reform movement. The association with St. F.X. Extension Department was a source of much encouragement to the people in the study clubs. They were always conscious of the prestige of a university directing their efforts.

The study club members learned from one another. Each club was likely to have a combination of young and not so young members, those experienced in the life of the community and the beginners. It is a striking fact that many of the charter members, directors and officers of the credit unions, were in their early twenties.

For example, Harold McKinnon, now president of Glace Bay Central Credit Union was in a study club at the age of fifteen. With two companions of his own age he faithfully attended the weekly meetings of a club consisting mostly of middle-aged men. He remembers that he did not talk very much, but that he learned a lot. He joined the credit union shortly after its organization and has served as a director and officer for forty years.

The younger members of study clubs gained much from their association with the more mature members who had lived through hard times and were able to relate what they were learning about economic co-operation to their own experiences. There were real leaders among them.

This was particularly true of the clubs in the industrial areas. The study club leaders were men who had been through the turbulent twenties when struggles against wage restrictions and opposition to organized labour had taught them social theory the hard way. It is significant that the founders of the credit unions in these areas were often veterans of labour union meetings. They were used to the proper conduct of meeting and their experience was invaluable.

Coal miners and steelworkers were accustomed to discussing issues and idealogies. The periods of industrial strife had been marked by the preaching of radicals whose attacks on the structure of society had been carried on in public meetings, the press and in all the places where the workers met. The study clubs had some members, who although they did not have much formal education, had read widely and had developed strong ideas about economic and social systems. Some of these were Marxists and their zeal for the tenets of communism was tempered in the free discussion of the clubs and the presentation of economic co-operation as an alternative.

A striking example of how study clubs changed the course of a life is the case of A.S. MacIntyre who was among the first to join a study club in Glace Bay. MacIntyre had had a stormy career in the labour movement. He freely admitted his leaning toward communism during his years as a member and an officer of the United Mine Workers. He had suffered much for his radical views both inside and outside the union.

The program of the St. F.X. Extension had a special appeal for him. In co-operative philosophy he found a substitute for Marx and he pursued the goals of co-operation with the same zeal that had marked his earlier quest for ways to help the working men or the "common people" as he called them.

A.S. MacIntyre was a born organizer and leader. He became the first field worker for St. F.X. Extension and the first president of the Nova Scotia Credit Union League. He served as managing director of the Nova Scotia Co-operative Union and was one of the co-operative movement's most prominent leaders in the Maritimes and across Canada. St. F.X. awarded an honorary degree to A.S. MacIntyre in recognition of his outstanding work.

How does one evaluate the effect of the study clubs on the lives of the participants? Dr. Coady often said that those who took part in the adult education program would never be the same again. Faithful study club members acknowledged the truth of this statement. Some were motivated to continue their formal education, even to making it through university. Others found that their experience in the clubs gave them enough confidence to help them to succeed in the occupations that were open to them at the close of the depression. Others rose to responsible positions in the labour movement. And the study clubs discovered the directors and managers of the co-operative institutions that were started in all parts of the Maritimes. The study clubs also produced leaders who became prominent in the co-operative movement throughout Canada.

Dr. Coady predicted that study clubs would discover the leaders need for a new era of economic reform. The leaders were out there among the people but there had to be some means of finding them. This was to be the unique function of the study clubs. "The lofty peaks," said Dr. Coady, "do not rise out of the level plain but out of the foothills."

From the regular study club meetings came the ideas that dominated more casual gatherings where the people were accustomed to talk about the economic problems of the times. These casual meetings were what Dr. Tompkins called "cracker barrel" education.

At a Rural and Industrial Conference in 1936, Dr. Tompkins gave one of the most dramatic speeches ever heard on the Antigonish movement. In part he said: "Our experience in the Antigonish movement is that there is more real education at the pitheads, down in the mines, out among the fishermen's wharves and wherever farmers sit and talk in the evenings than you can get from one hundred thousand dollars' worth of fossilized education (1936 prices!). It springs from the hearts and pains of the people. It is spontaneous. It is vibrant with motivation and motivation is the key to learning.

Between the formal opiate type of education and the spontaneous — call it the cracker barrel — I vote for the cracker barrel. The former does not fill any empty pantry; it does not bring milk and health back to babies already blighted by malnutrition in their toddling years. *We want ideas with marrow in them.*"

No matter what led people to join study clubs those who remained did so because the ideas discussed therein had marrow in them.

Chapter IV
MEETINGS

Those of us whose chief duty was to assist at the many meetings which were required in the educational program were the field workers. In the field were the people with whom we spent most of our time, giving talks, answering questions and helping to organize the credit unions and the other co-operatives.

The program was introduced to a community by the general meeting. This introductory meeting was Dr. Coady's special forum. The new interest in learning which led to the forming of the study clubs, the credit unions and co-operatives happened because Dr. Coady passed that way. When A.B. MacDonald followed close behind Dr. Coady the basis for "mobilizing" the people was firmly laid.

In the rural areas the meetings were held in the schoolhouse or the parish or community hall. The schoolhouse was a cheerless place after dark. There were only two occasions in the year when the school was lighted at night. One was the annual meeting of the school trustees and the light from a single lantern was sufficient for that event. The other was the annual Christmas concert when parents would improvise suitable lighting for angels and Santa Claus. The Extension meeting was held by the light of a few lanterns supplies by the organizers.

In late September and October it was chilly in the schoolhouse when the sun went down. Later in the season a roaring fire in the stove warmed up the building, sometimes too much for those who occupied seats nearby. Seating was not very comfortable. It was with some difficulty that a six-footer managed to fit his frame in the seat occupied during the day by his ten year old son. Latecomers might have to sit on the floor.

The halls were more suitable for meetings. They were better lighted and there were enough seats. But sometimes the halls were very cold. I was once cautioned by a woman who observed me removing my overshoes preparatory to mounting the platform. "Don't take them off," she said. "Your feet will be cold before this meeting ends."

The halls had another distinct advantage. There was always a kitchen where tea could be prepared to fortify the speakers for the long ride home. I do remember, however, that by the exercise of some ingenuity, tea could be brewed on the schoolhouse stove.

After the introductory meeting at which the study clubs were formed the people met in their small groups in the homes. After a period of study club meetings there was a re-assembling in joint meetings and the people were back in the schoolhouse or hall. From this assembly of study clubs there arose the organization known as the Associated Study Clubs. The name was invented by A.S. MacIntyre and the association was a logical development in the program. The club members who met weekly all studied the same questions, read the same material and looked forward to the same joint undertaking and the same practical results from their participation in the program. It was a natural outcome of this experience that they should come together to compare the work done in the weekly study sessions and to plan group action.

The Associated Study Clubs became formally organized with an executive consisting of a president, a vice-president, a secretary, and a treasurer. The duties of the treasurer were not very heavy as he counted the nickels and dimes that came into his coffers. The executive made all the arrangements for the monthly meeting which was held on a fixed date. At this meeting reports were given by the study club leaders, correspondence from the Extension office was received and literature was distributed.

The Associated Study Clubs organization enabled the field workers to keep in touch with the work of the clubs. In the beginning a field worker could attend numerous study club meetings. It was possible to visit several clubs in one evening but as the number of clubs increased this became impossible. Individual attention to clubs that wanted help could be given at the regular meeting of the Associated Study Clubs. A conference of the club leaders would be held before the opening of the general meeting. Questions raised by a club could often provide a way of giving useful information to all.

The Associated Study Clubs proved to be a wonderful invention for club members and field workers alike. The field workers had the schedule of the monthly meetings of the Associated Study Clubs in all the communities which they served. It was therefore possible to arrange for visits in advance and to make the rounds of meetings in an orderly way. The Associated Study Clubs meetings provided an effective way of introducing visitors who came to observe the St. F.X. Extension program.

The Associated Study Clubs meetings provided an enjoyable "evening out" at a time when there were few places to which the people could go for entertainment. They were pleasant gatherings where friends greeted one another. It was at these meetings that the Extension Department's public speaking contests took place. Young and old took part in these contests, thus gaining valuable experience for themselves and providing information and stimulation for club members. Very often the Associated Study Clubs meetings ended with entertainment in the form of music, singing and dancing in which visitors joined with enthusiasm.

The whole of industrial Cape Breton was organized in Associated Study Clubs, one organization for each town or district. Once a year they all gathered for a monster picnic on Dominion Day. This was no small feat to arrange with each organization responsible for part of the program.

The most important function of the Associated Study Clubs was to prepare the way for organizing credit unions and other co-operatives. When the decision was made to start a credit union the Associated Study Clubs meeting became the forum where all the questions were answered and all the details were planned. It was the Associated Study Clubs meeting that appointed a provisional board of directors, charged with the duty of obtaining a charter, drawing up by-laws and planning the first annual meeting of the new credit union.

As the number of credit unions and co-operatives increased so did the meetings requiring the attention of field workers. There were semi-annual and annual meetings at which speakers were required. At these meetings the topics were more specialized because members wanted to hear from somebody who could analyze and explain financial statements, report on legislation affecting co-operatives, and give advice on the operation of the business. These meetings also heard talks on co-operative principles and consumer education.

Then came conferences of all kinds. There were conferences of directors on a regional basis, store managers' conferences, handicraft conferences and educational meetings. Occasionally there was a staff conference at Antigonish. Although it was called a staff meeting it was attended by many other persons, including agricultural representatives, faculty members, and leaders in co-operative organizations. These were excellent meetings where fresh ideas came from each group. At the conclusion Dr. Coady would fire off some intellectual dynamite to send us home with the renewed zeal to make the rounds of the meetings again.

At intervals, for the purpose of generating new enthusiasm in the study clubs, the Extension Department would hold a "rally." This was an assembly of all the study clubs in a selected area. The most memorable of these rallies was held at Middle River, Victoria County, September 1940. Delegates came from all the Associated Study Clubs in Cape Breton, the credit unions and co-operatives and from the United Mine Workers and the United Steelworkers. The delegates long remembered the speech Dr. Coady gave on that day, one which for sheer power he seldom surpassed. He said so himself. While the old schoolhouse which was the scene of the rally remained standing, for many years after the delegates who had attended seldom passed by without referring to the day of the "Big Speech."

The day of the Big Speech followed a remarkable meeting of Phalen Local of the U.M.W. which Dr. Coady had been invited to address. The miners who called at the Extension office in the days prior to the meeting reported that there were expectations of a good attendance and some excitement among the local's members. There was a rumour that the few remaining members of the communist element in the local were planning to attend for the purpose of confounding the speaker.

Dr. Coady was accompanied by A.S. MacIntyre who, during his career in the U.M.W., had been a member of Phalen Local. The meeting was a smash hit. Dr. Coady got a thundering ovation. There was no heckling. Nobody dared. The radicals said not a word. Then Dr. Coady received what must have been the most unusual award of his career. By a unanimous vote the local passed a resolution granting him the privilege of attending any meeting of Phalen Local at any time he chose, with the privilege of taking part in the proceedings, like a member.

In later years when the mine in New Aberdeen was closed, Phalen Lcoal was disbanded. If the minutes of that famous meeting could be found, the record of that resolution would be an interesting document in a biography of Dr. Coady.

Chapter V
SOMETHING TO READ

The Extension Department supplied pamphlets, mimeographed lessons and books for the large number of students in the adult education program.

Kay Thompson, who was the first staff member to join Dr. Coady and A.B. MacDonald, was responsible for most of the study material that was used in the clubs. It was an age of pamphlets. Kay Thompson and Sister Marie Michael gathered pamphlets from far and near, from other university extension departments, from government departments of agriculture and from other agencies in Canada and the United States. No source of information escaped their diligent search. In answer to requests from the study clubs, the list of pamphlets grew to cover a wide range of topics. Requests were filled for such diverse subjects as the making of windmills and the tanning of leather.

The first pamphlets used in the clubs were those that treated the history and philosophy of the movement and the method of organizing credit unions and co-operative societies. These special pamphlets were written by staff members and were concerned directly with the group action to be undertaken by the study club members. If a particular subject required a pamphlet, somebody on the staff promptly wrote one. "How We Came To Be As We Are" which treated the development of the economic system from the time of the Industrial Revolution was one of the first pamphlets produced by the Extension Department. A handbook on the conduct of study clubs was provided to every club and it went through several revisions as the clubs progressed.

The pamphlet: "Credit Unions" by Joe MacIsaac became a standard text for groups planning to form a credit union. It was written in a question and answer form and soon became known as the "Credit Union Catechism." Thousands of copies of this pamphlet were distributed across Canada as one English speaking province after another followed Nova Scotia's example in enacting credit union legislation. (Quebec's legislation had been introduced by Alphonse Desjardins in 1906.)

"Maritime Techniques in Consumer Co-operation" by Kay Thompson included a short history of the Industrial Revolution in England, the story of the Rochdale Pioneers, an explanation of the Rochdale principles of co-operation and detailed instructions to be followed in the organization of a co-operative store. This pamphlet was required reading for all groups preparing for organization. It was translated into French and distributed for use in the Acadian communities of New Brunswick.

The pamphlet form was used also to record inspirational speeches given at the Rural and Industrial Conferences. These conferences were addressed by co-operative leaders and by recognized Canadian and American authorities in social theory. The principal addresses were printed and distributed to the study clubs and to individual inquirers. Reports of provincial and regional co-operative conferences were also made available to the study clubs.

Hundreds of articles from periodicals and lessons prepared by the Extension staff were reproduced by the women in the Extension office for use in the study clubs. The amount of work required was staggering by today's standards. The efficient copiers in today's offices had not yet been invented. The reproduction was made by hand on a mimeograph. Once Kay Thompson attached a note to a huge box of mimeographed lessons sent to a short course. It read: "I had to come up for air a few times while I was buried in this mountain of paper."

Amid the profusion of reading matter there was one unifying factor — the *Extension Bulletin* and later, its successor, *The Maritime Co-operator*. The *Bulletin*, established in 1933 and published by the Extension Department was the most important piece of study material used in the clubs. It kept the members of the clubs informed and interested in the philosophy and objectives of the movement. While the people in the clubs were occupied with the work of starting credit unions and co-operative stores the Bulletin helped them to understand that their efforts were an important contribution to the plan to reform social and economic conditions.

The Bulletin carried news from the study clubs, the new credit unions and the co-operative stores. The land, the industrial scene, the news of the spread of the co-operative message, the responsibilities and problems of women in the home were all subjects for the *Bulletin*.

In 1936, George Boyle, a professional writer, was appointed editor. Dr. Coady was convinced of the importance of the *Bulletin* in the educational program and he was particularly enthusiastic about George Boyle's appointment. On a tour of meetings he told his hearers that the Bulletin was entering a new phase. He would say: "The *Bulletin* is going to become so dazzling that you will have to put on smoked glasses to read it."

In 1939 the *Bulletin* was succeeded by the *Maritime Co-operator*. Until that time the Extension Department had paid the full cost of producing the *Bulletin*. It had always been expected that the co-operatives, when enough of them had been established, would assume the responsibility of paying for this periodical which provided their members with necessary co-operative education.

Ownership of the newly-named publication was transferred from the Extension Department to the organized co-operatives through an incorporated body: Maritime Co-operative Printers. A big effort was made to secure the support of the co-operatives. The subscription price was one dollar a year. Ray Cameron, a St. F.X. Extension staff member, was dispatched by A.B. MacDonald on a mission to visit all the co-operatives with the admonition to bring back three thousand one dollar bills.

In a brochure announcing the change in ownership, the *Maritime Co-operator* was described as the new organ of the co-operatives movement, that would bring together the common interest of labour, farmers and fishermen in the Maritimes. It was to have a section dedicated to the cause of labour. It would campaign for a revitalized rural economy and culture. It would furnish the driving force that would move people to build co-operative institutions. "It will fight the battle of the people on all fronts." In this sentence the announcement summarized what the *Maritime Co-operator* promised to do.

The *Maritime Co-operator*, through difficult times, kept the promise to fight the battle of the people on all fronts. This was its finest contribution to the co-operative movement and to the social and economic reforms that were the ultimate objectives of the Antigonish movement. Like the Extension *Bulletin* before it, it was the chief continuous means of education for the members of study clubs and co-operative societies.

The section on credit unions gave news of the growing movement and provided up-to-date information for directors and managers as credit union business increased in volume and complexity. The paper carried, as the times demanded, material on significant co-operative undertakings and on such problems as the taxation and the integration of co-operatives.

The *Maritime Co-operator* was as much an adult educational journal as a promoter of group action. It paid special attention to the economic problems of the vocational groups. Farmers read articles on such topics as depressed farm incomes, feed freight assistance, marketing boards, price supports and many others. The problems of steel and coal and of the organized fishermen were kept to the forefront. Among the issues of common interest to all groups were the economic disparity of the Atlantic region, the exploitation of buyers and borrowers, consumer education, interest rates, medical care and housing.

The *Maritime Co-operator* was especially dedicated to the promotion of harmony and unity among the vocational groups. It recognized that, to some extent, mistrust between labour and primary producers stemmed from misunderstanding or misinformation on the nature and cause of each other's problems. It made efforts to give information on the objectives and the difficulties of the various groups with the hope that knowledge would lessen the tension between them. This was the chief reason for the *Maritime Co-operator*'s labour section which, through the years, touched on every phase of the wage earner's life.

Wage standards, the cost of living, conditions in the work place, unemployment, industrial accidents, automation, labour legislation, exploitation, problems of the organized, — these are some of the topics that appeared in the *Maritime Co-operator*'s labour page. This section paid particular attention to producer-consumer relations and the farmer's cause was frequently upheld in the page addressed to the workingman's family.

Some of the topics on which the *Maritime Co-operator* informed and instructed its readers were considered controversial, depending on the viewpoint and interests of some readers, principally those outside of the co-operative movement. The paper was never afraid of controversy. To quote Kay Thompson Desjardins, who succeeded George Boyle as editor:

> "A paper that is too bland in content, too leery of offending those who are opposed to co-operatives or to needed social and economic change, would not be an asset to the movement. It would not provide the kind of stimulus that would move our readers to join co-operatives or strengthen their resolve to give continuous support to them, or to play their part in other worthwhile community endeavours."

The *Maritime Co-operator* practiced what it preached by following a restricted advertising policy. It did not advertise any of the products of its competitors, thereby passing up revenue which it badly needed to meet the costs of operation. For this reason the editorial policy was not restricted in any way. The editor could speak out fearlessly when a conflict arose between the interests of its competitors and those of the co-operative movement and consumers generally.

The *Maritime Co-operator* enjoyed a reputation for

excellence in the co-operative press and it was the recipient of numerous awards, to the publication itself and to individual contributors. Awards for excellence in writing came from the Canadian Women's Press Club, from the Co-operative Editorial Association and the Co-operative News Services of the United States.

The women in the St. F.X. Extension Department made a remarkable contribution to the success of the *Bulletin* and the *Maritime Co-operator*. Sister Marie Michael was the first editor of the home and family page of the *Bulletin*. She was followed in this assignment by Zita O'Hearn (Cameron) as home editor of the *Bulletin* and later for the *Maritime Co-operator* until 1970 when she became editor of the *Co-operator*. From 1942 until 1947 Kay Thompson Desjardins wrote the labour section until she succeeded George Boyle as editor. I wrote the labour section from 1947 until 1978 and a consumer education column for part of that period. During Kay Desjardin's twenty-three year term as editor the talented Mary MacIntyre McNeil wrote the credit union page and columns on a wide variety of topics until her death in 1972.

For a long period the *Maritime Co-operator* was written and managed entirely by women, a unique distinction in the co-operative press. During this time Dr. Coady was our greatest supporter and a regular columnist. He was proud of the *Maritime Co-operator* and boasted about the women who produced it.

Early in the thirties the Extension Department established an "open shelf" library in Antigonish. The service began with "travelling" boxes containing approximately fifty books and circulated from one community to another for the use of the study clubs. Borrowers could also obtain books by mail from the library, with the Extension Department paying the postage both ways.

Readers who are accustomed to borrowing books from the public library or from the bookmobile at the farm gate may find it hard to imagine a time when the only way to obtain a book was to buy it. That is how it was when there were no public libraries. Few, if any, of the study club members could afford to buy a book. Many families could not afford to subscribe to a daily newspaper. For some people getting books from the Extension library was like coming upon a hidden treasure.

In 1936 the Extension Department opened a small library at its branch office in Glace Bay as part of its service to the study clubs in industrial Cape Breton. New books were procured for the library and they were supplemented by books loaned by the Antigonish library and books that had been collected in the Glace Bay office prior to 1936.

At that time Dr. J.J. Tompkins was parish priest at Reserve. He was at the forefront of a campaign for the establishment of a system of regional libraries in Nova Scotia. With a grant from the Carnegie Foundation a regional library had been organized in Prince Edward Island. Norah Bateson, the director who had set up the P.E.I. library system, was a friend of St. F.X. Extension. She had been a guest speaker at a Rural and Industrial Conference where the principal discussion had been regional libraries. Following her P.E.I. assignment she had taken a position in Halifax and was available for consultation. Dr. Tompkins and Norah Bateson were kindred spirits, dedicated to the idea that libraries should be essential components of free public education.

The announcement that St. F.X. Extension would open a branch library in Glace Bay filled Dr. Tompkins with delight. He had already established his "People's" Library in Reserve. Every new venture in library service seemed to be a step forward in the campaign for a regional library.

I came to Glace Bay from Pictou County to do field work and found to my surprise that I was to be the librarian. It was not unusual for Extension workers to be assigned to jobs for which they had no training and which they had to learn to do by doing. A large number of boxes of books had preceded me and I had to catalogue them and make them ready for circulation. I knew nothing about organizing a library. I had never heard of the Dewey Decimal system and I was an indifferent typist. I sat in the midst of those books and despaired.

Dr. Tompkins promptly came to my rescue. He said that it was very easy to organize a library. "Nothing to it," said he, adding that his own library was run by a woman with a master's degree in library science. He offered her assistance and Sister Frances Dolores cheerfully undertook to make an instant librarian out of me. Encouraged by her advice and Mr. Dewey's invaluable book, I set to work.

The news that books could be borrowed from the Extension office spread rapidly. Dr. Coady, at a series of meetings, invited anybody to come a-borrowing. I was present at those meetings, fervently hoping that the readers would not come too soon. But the promotion by Dr. Coady and A.S. MacIntyre and the credit union directors brought results and borrowers came, to be told by a harrassed pseudo-librarian that they could not take out books until they had been properly catalogued.

The work proceeded slowly. It was a busy time in the study clubs and there were many meetings to attend. To try to speed the opening of the library to borrowers I brought books home on weekends and on those evenings that were free of meetings. These were books that needed close examination before they could be put in the proper class invented for them by Mr. Dewey.

This devotion to the library nearly brought about my incarceration in the town jail. The town's police chief stopped to arrest me as I walked, carrying an armload of new books. Acting on complaints from irate citizens he was on the lookout for a "strange woman" who was peddling "bad books."

Another incident concerning "bad books" led me to institute my own index of forbidden books. In my absence MacIntyre loaned a book, a widely-acclaimed novel which had just arrived and had not been catalogued, to a clergyman who was a steady borrower. Today the book would not raise an eyebrow but in 1936 it was shocking. After a reprimand from the parish priest, who was a friend of the reverend borrower, I consigned all doubtful books to a cupboard from which A.S. MacIntyre was firmly and permanently excluded. I had no time to soothe shocked readers. Soon I found that I could get valuable assistance from a few trusted borrowers who were happy to examine books from the forbidden cupboard and report on their suitability for a place on the shelves. My consultants were a great help and we received no more complaints.

Finally order was established and we soon had a list of regular readers. The library was open to everybody but it was particularly useful for study club members, credit union and co-op store members and directors. The library was housed in one of the two rooms that comprised the office for field work in industrial Cape Breton. There was no hushed library atmosphere there. Study club members came to get material for their clubs and to talk about their activities. Directors of credit unions came with their questions and reports. The library was the reception room for the many visitors who came to observe the study clubs in Cape Breton.

Typing had to be done in snatches between visitors and borrowers. It is a wonder that we managed to get so much work done with the facilities and office equipment we had.

With the library system established in Glace Bay we began to promote its use in the surrounding areas. This we did by providing small collections of books to be rotated from one community to another. It was a sort of mini-regional library system and it was a time-consuming task. Eight communities received books under this plan. Each community had a library committee which was responsible for providing a place from where the books could be distributed and for finding a librarian who would take charge of the books and keep the records. The most convenient place for this mini-library was the credit union office during business hours.

The library committees raised money locally and made some contribution to the upkeep of the library.

The plan worked fairly well but it was found to be too cumbersome. It was difficult to keep the librarians interested because the collections of books were just too small. There were not enough books to permit the wide range of choices necessary to cope with the demand from borrowers. If we sent 100 books to a community there could not be more than one or two in a particular category. Borrowers needed a wider selection from which to choose.

In spite of its limitation our mini-library system with its small collections of books produced good results. It stimulated interest in the campaign for a regional library. Norah Bateson was called from Halifax to speak at meetings, including one at Phalen Local of the United Mine Workers.

She was impressed with the enthusiasm of the miners who were in the campaign to obtain a regional library. Credit union and co-operative joined the campaign.

Norah Bateson and Fr. Tompkins prepared the brief to be presented to the annual meeting of the Joint Expenditure Board of Cape Breton County. This was the first formal request from citizens to the Board which had the authority to set up a regional library. Everybody was hopeful. The brief was supported by many organizations and the cost per capita for the proposed library plan was a small fraction of the total expenditure for education in Cape Breton County.

There was disappointment when the Joint Expenditure Board turned down the proposal. There was gloom everywhere. Norah Bateson was astounded to find that the administrators of public funds for education would refuse to provide libraries as part of the educational system.

Dr. Tompkins was irate. He summed up the short-sightedness of the authorities in this way: "A society which pays the cost of an educational system that teaches people to read but does not provide books is like a railway that provides trains but saves expense by eliminating timetables. *Nobody knows when the train goes.*"

Eventually a system of regional libraries did come to Nova Scotia but there is no doubt that its coming was hastened by the determination of the people in the study clubs and the efforts of the St. F.X. Extension Department.

Chapter VI
THE SHORT COURSE

From the experience gained in the supervision of the study clubs there arose a new technique in the adult education program. This was the short course which was first tried in 1933 and which reached its highest development in the forties. The communities needed leaders to take responsibility for developing local co-operatives, and, as the number of credit unions and co-operatives grew, it soon became evident that some specialized training was necessary for directors, officers and managers.

The first short courses were held at St. F.X. They were month long courses, held during the winter because that was the most convenient time for students from the rural and fishing areas. It was not, however, the most convenient time for travelling. Alex MacIsaac, who in later years became managing director of the Nova Scotia Co-operative Union and a minister in the provincial government, often amused his audiences with the description of his walk through snow banks all the way from Giant's Lake to Antigonish to attend the short course.

When the students were safely established in Antigonish they had little worry about how they were to reach home in four weeks' time. The course was interesting and exciting. Let it snow!

The program for the short course was a crowded one with classes in the philosophy of the Antigonish movement, basic economics, co-operative principles, and bookkeeping for credit union and co-operative employees.

The course included practice in conducting study clubs. These were lively sessions in which students took turns as leaders to prepare themselves for the work they expected to do on their return home. The students were also introduced to other forms of discussion as in a panel or a forum. They had access to a wide selection of books from the Extension library and they could choose, from the large quantity of mimeographed material and pamphlets, something that they could use in the local study clubs. And the highlight of any day's events was the frequent appearance of Dr. Coady to fire the students with the zeal to go back to their communities to do great things.

Organizations in the local communities showed a deep interest in the short course. The members were mindful of their need for trained leaders to help in the development of their credit unions and co-operatives, and they used many ingenious methods to raise money to pay for deserving students, none of whom could finance a month at the university. Many a card party and dance preceded the opening of a course. These fund raising activities were usually successful because the courses were always filled with eager students who were thrilled with the opportunity to go to college for a month.

The directors of Coady Credit Union met one winter evening to discuss how to finance for the short course a volunteer teller who showed a particular aptitude in bookkeeping. They decided that winter was not a good time to stage money-raising functions, so the directors themselves took out a loan which they would repay by summertime activities. The scheme worked fine.

It was not difficult to find students to attend the courses. They who were selected were pleased with the prospect of attending St. F.X., the source of the new education which was making a change in the life of the community. The list of students who attended the short courses includes the names of many who later became leaders in the co-operative and labour movements in the Maritimes and in other parts of Canada.

The St. F.X. short course continued to be an annual event until the end of the thirties when changing economic conditions brought about by the war made it difficult for students to attend a four week course. It was then that the short course went to the community. This had to be a very short course because few people could spent more than a few days away from work, during this period of full employment. In the industrial areas the course might take the form of a series of evening classes. In the rural communities it was possible to hold two or three day courses.

The community course was sometimes a gathering of a particular group. The directors of the co-operatives and the credit unions in a given area would be brought together for a day-long course, and a general meeting would be open to the public in the evening. Handicraft and nutrition conferences were organized for women who were active in the study clubs. These conferences were open to everybody and new members were sometimes brought into the study clubs in this way.

During the forties short courses began to replace the study clubs which had slowly declined in numbers. Various reasons could be given for this decline. While the clubs were engaged in the work preparatory to the organization of a credit union or a co-operative store there was a compelling reason for full attendance at the study club meeting. When these co-operative enterprises were finally established, interest in the club meetings was likely to diminish. The clubs became more specialized and of particular interest to a smaller group, consisting of directors and committee members. They had been the most active members in the study clubs and were now responsible for the operation of the co-ops that had resulted from the study program. They were required to attend many meetings associated with their new responsibilities. These directors were well aware of the need for continuing education among co-operative members and the decline of the study clubs was a source of much concern to them. The records of the early credit unions show to what extent the problem of member education occupied their attention.

Another reason for the decline of the study clubs was the sudden change in economic conditions resulting from the outbreak of war. For example, coal miners suddenly found themselves on full employment for the first time in many years. The young men rushed to join the armed forces. Very soon Canada's fuel requirements became so critical that the federal government "froze" the miners in the pits and no more enlistment was permitted. But the loss of so many young miners was serious and the men who remained worked very hard. Their performance in the war effort was exemplary. They worked six days a week, often with overtime, and they "hung up" records. There was little time for study clubs.

The changes in the employment situation affected everybody and there was less inclination to give up evenings to study club meetings. The short course grew in importance and it can truly be said that in the forties the short course to a great extent replaced the study club in the St. F.X. Extension movement.

A particularly busy period in short course activity began when in March 1939 the federal government granted St. F.X. funds to carry on educational work in the fishing communities of the Maritimes. By the time the grant was received St. F.X. had experience with the short course and that was the method chosen for the new program in the fishing areas.

As a member of the short course teaching team I travelled to communities throughout the three provinces. Courses were held in most of the fishing communities in Nova Scotia. We covered Prince Edward Island thoroughly and the names come quickly to memory: Mt. Stewart, North Rustico, O'Leary, Morell, Tignish, Wellington, Georgetown, Souris, and others from one end of the Island to the other. The director of Extension for St. Dunstan's University and frequently Dr. Murphy, the president, accompanied us on these courses.

In New Brunswick the program was assisted by St. Thomas College and St. Joseph's College (Memramcook) and Sacred Heart College at Bathurst. Besides courses held at these colleges, others were held in many fishing communities: Shediac, Buctouche, Dalhousie, Tracadia, Escuminac, Tabusintac, to name a few.

After a few trials the short course became an efficiently run school-for-a-week. When a place was chosen for a course there had to be careful preparation by local leaders. From St. F.X. Extension these leaders received specific instructions on the required arrangements, but for many of the preparations, the local committees were on their own and they always responded well. Because the students came from adjoining areas to a central place transportation had to be provided morning and evening. Two meals had to be prepared each day because the proceedings extended to an evening session. There had to be a hall for classes and a place for serving meals. Living quarters for the teachers and speakers had to be found where there was no local inn.

The local leaders invariably showed a great deal of interest and enthusiasm. The people of the community seemed to be excited by the invasion of visitors and, by means of various committees, headed by the best go-getters in the place, everything went smoothly and preparations were always complete. Only on one occasion did the hall committee barely complete preparations. At Escuminac as the Extension car drove up to the hall at ten minutes before nine on Monday morning the local organizers were nailing the last board on the new front steps. One of the volunteer carpenters shifted his hammer to the left hand to extend his right in greeting.

The local clergymen were a great assistance in organizing the courses and they usually attended all the classes. Meals were prepared by a women's society where one existed. If not, there was a community meeting where plans were made to feed the coming co-operators. The meals were always excellent and served with great dispatch. In those days nobody counted calories and it was remarkable that none of us put on weight and some remained thin as rails. As for cholesterol, if the doctors had by that time discovered it, we remained in blissful ignorance of its menace to our health.

I particularly remember the preparations at North Rustico. There was a new school with a fine auditorium which had a good stage and a large screen where we could show films. The parish had recently organized the boy scouts who, resplendent in their new uniforms which they were wearing for the first time, had been assigned to wait on us and run all errands with speed and efficiency. They were such delightful little boys and so eager to do their good deeds that we invented little chores for them to carry out. All went merrily at North Rustico for the first three days until a snow storm halted the course before the big finale which had been planned for the closing night. It was a disappointment to all, teachers and students, that the course which had begun with such promise did not continue, as it had been planned, to a triumphant end.

The short course team, headed by A.B. MacDonald would arrive in a community on Sunday afternoon, with boxes of pamphlets and mimeographed lessons, charts, posters, projector, films, screen, typewriter, boxes of demonstration articles for consumer education classes, books, and personal baggage. As we unloaded all this stuff we looked like peddlars ready to sell their wares.

The classes began on Monday morning at nine o'clock and continued until five in the afternoon with a break for the noon meal and a short recess mid-morning and afternoon. There was nothing to eat or drink at these recesses. The coffee break had not yet been invented. The time table was carefully prepared and included instruction in group marketing for fishermen and farmers, agricultural topics, principles of co-operation, credit union practices, problems of consumer co-operatives. It had to be a diversified program because the people who came to the courses had many questions on all these topics.

Fishermen had their own special problems relating to processing and marketing. But fishing was not a year-round occupation and part of the fisherman's annual income was derived from the land, and so there were speakers from the departments of agriculture. Since we stayed in the classroom throughout all the seasons, we learned a lot about lamb, eggs and potatoes.

All of the students belonged to credit unions or co-operative stores or both and many of them were directors and officers. There were women at these courses, all interested in the development of the credit unions and the co-op stores in their communities. There were young people out of school but not old enough to join the armed forces.

Part of the day was taken up with practice study clubs. These were serious sessions which accomplished two purposes.

They gave the students practice in discussion and they provided a forum for a review and evaluation of the performance of the credit unions and co-ops in the communities from which the short course students came. These model study clubs were made up of participants from each community represented at the course. We purposely arranged the clubs in this way so that there could be an exchange of reports on the success or failure of performance by the co-ops in different areas. The clubs were given pertinent questions to discuss and a time limit placed on the proceedings. Reports from the individual clubs were then considered at the general assembly.

The study club technique was an effective way of finding out how the local credit unions and co-ops were progressing and what problems they were encountering. Because each club was required to come to a consensus on the questions that had been assigned, the plenary session was usually satisfactory both to the students and teachers.

By five o'clock the day class would end and the students would gather for a meal and a break before the evening session which began at seven o'clock. There was no time for loafing that week. The evening meeting was open to everybody and literally everybody came. The people who did not attend the day classes observed the commotion in the place and were curious to find out what was going on. The women who had been hard at work preparing and serving meals did a speedy job of washing the dishes, hurrying home to feed the family, gathering up the children, changing their aprons for dressier clothing and getting to the hall with a few minutes to spare.

The children were excited because they expected to see a film. In most of the places where the courses were held the children had never before seen moving pictures. They would sit patiently through a lecture in anticipation of the movie to come. This was usually a film on a co-operative subject, not well suited to the smaller children, but they enjoyed it nonetheless. We had the film, "Here is Tomorrow." At one dramatic point a whole screenful of baby chicks were seen emerging from their shells. The children were enchanted with that film with its chickens and tractors and cows and fields of growing things. They loved it all and were models of good behaviour.

At these evening sessions we took turns with the principal speech. A.B. had the first meeting with a complete outline of the program for social change. I had the second with a demonstration and talk on consumer education. The third evening was devoted to the credit union with a special speaker who would come for that meeting. Speakers from the Departments of Agriculture and Marketing were available for the next session. Rev. J.D. Nelson MacDonald, pioneer co-operator and member of the St. F.X. Extension staff, was a prized addition to the roster of speakers for the short courses. When he could be captured for an evening meeting the program was considered special indeed. And if we were lucky, Dr. Coady came to give one of his spell-binding speeches, a "rip-snorter" as he called it. The evening meetings ended with music and dancing and a good lunch.

The short course was a valuable educational method because it provided sufficient time for a thorough discussion of local problems in co-operative enterprises and for the exposition of the whole program of adult education. Between classes and during free time the inspector of credit unions and co-operatives, and the agricultural representatives, were available to talk with individual students who had specific questions for which they sought answers.

And, best of all, for the students who gave up their work for a week to attend, the short course had A.B. MacDonald, the head and front of the enterprise, whose energy, skill and dedication made the short course a success wherever it was held. He was eminently qualified for the task. As the associate director of the St. F.X. Extension Department he was an able exponent of its program.

A.B. was an expert in the principles and practices of the co-ops and credit unions. He knew well almost every credit union and co-op represented at the courses, having assisted at the organizing of most of them and in the preparation of the legislation under which they were formed. He could talk with the farmers as an expert because he had a degree in agriculture and had practical experience as an agricultural representative. He was a first-class trouble-shooter who was often called upon to settle disputes, something which he accomplished smoothly or as he put it himself, "with an olive branch in my hand." He was a superb teacher. He used to say that the best time of his life had been the years that he spent in the classrooms as a school inspector. It was because of his love of teaching that he probably enjoyed the short course more than any other phase of his busy life with its wide variety of tasks. Added to these qualifications were his friendly manner and his sense of humour which expressed itself in a fund of hilarious stories with which he prepared his audiences for the important messages in his talks. A.B. left a community better because he had passed through.

The chief disadvantage of the short courses to the travelling teachers was that they took place in the winter with all the problems that season brought. There was never any problem getting to the site of a course, but getting back, particularly from Prince Edward Island, was always a source of worry. Yet, in spite of snow, ice, wind and tides, I have happy memories of the courses.

I particularly remember the course held in Wellington, P.E.I. because the weather was perfect that week for going, staying and coming. The winter roads were at their best for sleighing. I stayed at a home which must have been the hospitality centre of that community. I was amazed at the number of teams of horses that evidently could be accommodated in the barn. In the hour preceding the evening meeting, as I sat resting before the big event, I heard over and over again the cheery greeting as, one after another, the sleighs rode into the yard: "Boys, get the lantern and put the horse in the barn."

Dr. Coady was at Wellington and at his very best. He was enjoying the winter weather with the snow, the sleigh bells, the frosty air to make us all strong and the full moon shining splendidly just for the co-operators.

On the Friday afternoons we parted from our students as friends. We talked so much during a course in the classes, before and after evening meetings, during meals, during breaks, that once a student commented that it was a good thing that tongues don't get tired. But sometimes on a dreary bus trip late at night on the unpaved road from the Strait of Canso to Sydney it seemed that weariness did extend to the tongue too.

Chapter VII
A POCKET IN A SHIRT

The credit unions that were founded in the thirties are, one by one, celebrating their fiftieth anniversary. Not many of the charter members are living, but those who have survived the half century must inevitably recall, during jubilee festivities, the dismal quarters where the movement started a half century ago. In moments of enthusiasm the study club members, planning to found a credit union, had visions of rivalling the chartered banks, but most of the time all their attention was centered on the work at hand, the practical preliminary work of getting organized.

It was to be expected that the credit union would be the first co-operative to arise from the study clubs. The idea was first proposed by Dr. Coady at meetings where he advocated the credit union as the fundamental form of group action that would give the people control of their own finances. The men and women of the study clubs were not likely to forget how Dr. Coady had destroyed the myth that the banks were shrouded in obscurity, and that they were hallowed places which the poor man enters timidly, cap in hand, to appeal to the kindness of the banker for a loan. He had made it clear that banks do not lend their own money. In this short sentence: "Banks have money to lend because people put money in the banks," the myth was destroyed.

Thereafter the rapid growth of the credit unions happened for two reasons. The first was that the credit union was the easiest co-operative to organize. "Do the evident, feasible things first. Begin where you are," Dr. Coady had advised. The credit union was evidently feasible, well within the grasp of the members of the study clubs. No matter how poor they were they could make a start.

The second reason for the growth of the credit union movement was the urgent need for an acceptable system of credit in both the rural and the industrial areas. The men and women in the study clubs were well acquainted with the problems of debt. To understand fully the conditions that motivated the founders of the credit unions it may be useful to compare the credit needs of people fifty years ago with the role of credit for economic survival and not merely as a convenience. A 1984 credit card holder may simply defer the payment of a bill because it is convenient for him to do so, not because he is unable to pay cash at the moment of purchase of goods or services. In pre-credit union time people needed credit because they had no cash at all. They absolutely *had* to defer payment until cash became available and, in the meantime, they needed credit. This almost total dependence on credit kept some people in a sort of slavery to those who "carried" them.

In rural districts farmers had to delay payment until they could sell their products. They needed credit to buy household necessities as well as long-term credit for farm equipment. Fishermen lived under a system that kept them in perpetual debt. On one occasion at a meeting a fisherman recalled a "good" year when there was a balance in his favour with the merchant who bought his fish. He was able to spend the balance in the purchase of a rocking chair, longed for by his wife for the purpose of rocking the baby to sleep. The cost of the chair used up the whole balance.

In the industrial area where the credit unions developed rapidly the workers had been accustomed to a similar dependence on credit. Many of the men in the study clubs had had experience with the credit system in the coal company stores, which had kept miners' families in debt with no hope of emerging from that undesirable state. Reaction against the system had led Cape Breton miners to destroy the company stores during the strike of 1925. Men in the study clubs who had been through the upheaval of 1925 were the first to admit that getting rid of the stores had not solved the problems of the miners who needed credit. The credit union, an alternative system based on equity, democratic control and active participation by the members, was an exciting proposal. It is easy to understand why the credit union appealed so strongly to students in the clubs in both county and town.

Members of these clubs took up the study of the credit union with much enthusiasm. They had no fear of failing to make bankers out of themselves. They found inspiration in the history of the credit union movement in Germany, the United States and the caisses popularies in Quebec. The names of the credit. union heroes, Raffeisen, Filene, Desjardins, became familiar. The eager students learned how ordinary working people had become bankers, saved their money in their own institutions and established systems of credit which they themselves controlled.

Some impressive statistics were used at the meetings that studied how to found a credit union. In the United States after the market crash of 1929 five thousand banks had failed, but not one credit union had suffered the same fate. I cannot remember that we followed the trend in the misfortune of the banks in the remaining years of the depression. Evidently we thought that five thousand was a good enough figure in the case of failing chartered banks and we saw no need of pursuing the point any further. It made an interesting addition to a talk on the soundness of credit unions and never failed to impress the audience of potential credit union members.

The pamphlet, "Credit Unions" written by Joseph MacIsaac and published by the Extension Department, became the standard text in the study clubs. It was made up of questions and answers and the questions that MacIsaac had not thought of — and there were few in this thorough presentation — were asked at Associated Study Clubs meetings where field workers were present and prepared to answer. The chief speaker in this undertaking was A.B. MacDonald who travelled to every community where meetings were held in preparation for organizing a credit union.

The preparation proceeded on two fronts. While the people were studying Joe MacIsaac's book and every other piece of material on credit unions they could find, the Extension Department was at work on the legislation required for the organizing of a credit union. When R.F. Bergengren, an authority on credit unions, came from the U.S. to assist in the preparation of legislation, excitement ran high in the study clubs. The credit union legislation for Nova Scotia was enacted in 1932 and the clubs were ready.

As the clubs pursued the study of the credit union, the members began to form a new opinion on the nature of credit. They understood that the providers of credit did not confer any favour on the borrower, and for the first time, they began to talk about the price the borrower had to pay when he asked for "time" to pay a bill. The loan at the bank and the credit at the store were no longer considered a privilege to the borrower or customer but instead a lucrative transaction for the lender. The myth surrounding the banks was completely shattered.

But, just like the banks operating in the community or in the town a few miles away, the credit union could loan money only if people put money in it. The people in the clubs were convinced that they could run a credit union successfully if they had the money. The problem was that nobody had any money and there was little hope of acquiring any savings from the meagre income from the land, the sea or the labour force.

The fact was inescapable. If they were to start a credit union they would have to save the money somehow. At first this seemed impossible. A miner, leaving a meeting in Glace Bay where an Extension speaker had urged his hearers to start a credit union, said to his companions: "How does that fellow expect us to start a bank? We did not have ten cents among us in the hall."

The Extension workers then concentrated their efforts in persuading the people that they *could* save. There was a spirited campaign on the merits and the possibility of saving. The exhortations made sense. A credit union could be run no matter how small the savings. What did it matter if only fifty dollars were in the fund? That amount could be loaned while the manager waited for more to come in.

Nobody taught the wisdom of saving better than A.B. MacDonald and he had a number of hilarious stories by way of illustration. His favourite was about Roddie and Archie. According to A.B.'s yarn, Roddie used to go to Sydney from Bras d'Or every Saturday to sell vegetables. On his way home he always stopped at Mary's where he would procure a bottle of the good stuff which Mary dispensed illegally. On Saturday night Archie would come to visit and the bottle would be passed around. One Saturday Archie came over as usual but he waited in vain for the appearance of the bottle. After a full hour had passed with no sign of refreshment, Archie ventured to ask: "And did you stop at Mary's?"

"That I did," replied Roddie. After a few moments he continued: "But there was nothing there. The law caught up with Mary. But I left the money there just the same."

"And why did you do that?" asked Archie.

"I was afraid," replied Archie, "that if I took the money home, I would spend it on some foolishment."

The moral: Archie needed a credit union to keep his money safe from the temptations of foolishment.

Once the study club members were convinced that saving was possible some of them, particularly those in the industrial areas, set up a saving plan in anticipation of the opening of their credit union. The secretary of the club would receive the savings which usually amounted to ten cents per week and carefully record them. A good week for the coal miners would allow for twenty-five cents in the savings account. As the contents of the piggy bank grew so did the confidence of the members. They were well informed on the history of the Rochdale Pioneers who had saved a penny a week to gather capital for their co-operative store. In the credit union history of Nova Scotia the dimes and quarters of the poor have as important a place as the pennies of the pioneers of Rochdale have in the history of co-operative merchandising.

Along with the exhortation to save, the Extension program taught another important credit union principle. This was that every community had within itself the resources to take care of the ordinary short term needs of its people. Those who were planning a credit union had to believe this or the struggle with the dimes and quarters would be useless.

The community basis of the credit unions was important. It was A.B. MacDonald who first used the old saying: "as handy as a pocket in a shirt," to describe the convenience of the local credit union. After that we all used it. The members particularly liked being able to set the hours of business that best suited them. The community credit union was particularly well suited for saving in small amounts. A farmer would not likely hitch up the team to drive into town to deposit twenty-five cents, but this was easy to do in the credit union. For example, the Cheticamp Credit Union made its collections at first on Sunday morning before church services, every two weeks, at which time the members were committed to deposit the sum of ten cents. And the picture of the industrial worker getting dressed up to go to town to the bank to deposit twenty-five cents always brought forth much amusement.

Most of all the members liked the idea of becoming owners of this new kind of bank. Poor as they might be, there was a good feeling about friends owning something together, helping one another in a practical economic enterprise.

It was an interesting experience to have a part in the process of organizing a credit union from the time when it was just a study club topic until the opening day. The procedure followed a set pattern. The Associated Study Clubs would make the initial plans and a date would be set for the meeting to make an application for a charter. This meeting was always a memorable one. As one by one the charter members wrote their names on the application they were well aware that they were making history. They were beginning something that would help to right the errors in the Great Default.

Between this meeting and the arrival of the charter a provisional board busied itself making plans for the first annual meeting, looking for a place to do business, searching for a manager and for members to fill the positions on the board of directors and on the committees.

The arrival of the charter was an important event. Everything was ready for the first annual meeting where the constitution and by-laws would be adopted. Choosing a name for the new credit union was a pleasant duty but I remember one occasion when it was done with sadness. The Stellarton study clubs had applied for a charter and one of the most enthusiastic charter members was Alexander Beaton, who had been favoured to be the first president of the new credit union. But before the first annual meeting took place there was an explosion in the Allan Shaft in the Stellarton coal town and Alexander Beaton lost his life. It was only a few days after the new credit unionists buried their friend that at the first annual meeting they voted unanimously to call the new bank the Alexander Beaton Credit Union.

The first annual meeting elected the directors and the committee members. Eighteen people had to be found to assume the responsibility of operating the credit union. These were not on-the-spot elections. The work done in the study clubs assured that leaders would be found, willing to spend long hours without pay, in running the credit union.

The supervisory committee was the easiest to fill since some skill in arithmetic was the only qualification necessary. There was absolutely no office equipment. It would be years before the credit unions could afford adding machines or typewriters. If you were on the supervisory committee, you added, subtracted, multiplied and divided figures without any mechanical help, just as in your school days. It helped that at the beginning there was not much money to add up and that arithmetical skills could be expected to improve with experience as the business grew. It was fortunate that every credit union seemed to be blessed with some member who was "good at figures" and was willing to use this skill for the benefit of the organization.

The credit committee had a more difficult task to fill because its members constituted the jury that decided whether or not an applicant would receive a loan. The by-laws stated that a loan should be granted for a "provident or productive purpose" and it was the first duty of the credit committee to decide when the purpose of a loan application fell into one of these two categories. When the committee was satisfied on this point, it then had to consider the applicant's ability to repay the loan.

There was a limit on the amount that could be granted on a character loan. In a community where everybody knew everybody else the character of an applicant was common knowledge. But in the towns as the membership grew fast it sometimes became more difficult to make judgments on character loans and the credit committees spent much time in special meetings.

Endorsers were required for loans and the credit committe members probably knew all the endorsers as well as they knew the applicant. Directors were generous in endorsing loans although they were subject to more stringent rules than the other members. The safeguards were written in the by-laws. An applicantion made by a director had to be approved by what was called the full board — the directors, supervisory committee and credit committee sitting together. Ease in obtaining a loan was not a factor in a member's decision to stand for election to the board. There was no personal advantage to a member seeking office. The people who guided the credit unions in the early years gave their service freely and received no special privilege.

The great care with which the credit committees carried out their duty can be seen in the minutes of their meetings. Records of the Glace Bay Central Credit Union show that one of the directors, who later became its president, applied for a loan of twenty-seven dollars. Although he had twenty-eight dollars in the credit union, the full board met to pass the loan.

It is impossible to praise too highly the work done by the early credit committees. Their decisions could ensure the success of the organization or seriously impede its progress. The members were conscious of this when they chose their credit committees. During the period of credit union study much stress had been placed on the importance of making the proper choice for members of the credit committee. A.B. MacDonald had laid down the rule in many a meeting: "The credit committee must be composed of men with *'good sound judgment.'* "

The first directors' meeting, which followed immediately after the first annual meeting, elected the officers, appointed a manager, and made plans for the opening day. The manager had to be somebody with some experience or at the very least be a fast learner. Since there was no pay attached to the position the new manager had to have faith in the possibilities for the new bank. If he had a regular job, he would have to give up his free time to attend to the business of the credit union.

Some of these early managers stayed with the credit unions long enough to become part-time employees with pay or full-time managers. But most of the credit unions that are successfully doing business today exist because their first managers worked without any remuneration. They were rewarded only by the satisfaction they got from helping to build something in which they firmly believed and for which they had great hopes.

Opening day, a time for much rejoicing, came. Equipped with a ledger, pass books, deposit slips and a pencil, the manager took his place behind a table to serve the members who came to make their first deposits. These few items constituted the total amount of the credit union's assets on opening day. The bill for the supplies was most likely still unpaid. Money for this purpose would be raised by the directors who would organize card parties with an admission fee of ten cents, the proceeds to be applied to office supplies for the credit union.

Those members who had saved five dollars in the study clubs would pay this amount for a share and deposit twenty-five cents as an instalment on a new share. All the directors and committee members would be there, greeting members, answering questions and rejoicing in the final achievement of the project for which they had been so long and carefully preparing.

My memory of opening days is that the spirits of the members were high but that the surroundings were gloomy. The temporary "office" was usually dark and dingy, not to be disdained, however, because it was rent free, obtained through somebody's generosity. In the towns the members on their way to the opening might have passed a fine looking bank with more money in its safe than the credit union could gather up in a very long time.

The proud credit unionists did not like to think about *their* money going into that bank safe on the next Monday morning. It seemed like a traitorous act, but where else to put this money since the credit union could not afford a safe?

At the first annual meeting of the New Glasgow Credit Union the time came to select a bank in which the credit union would deposit its weekly collection. After a short discussion a motion was passed designating the Royal Bank of Canada. The chairman, who was noted for his wit, brought down the gavel. "Motion carried," he said, "I'm sure Sir Herbert Holt will be very pleased." Sir Herbert was president of the Royal Bank of Canada. The New Glasgow Credit Union did not yet have a cent to deposit.

Opening day was a day for dreamers. There was no ribbon cutting, no ceremony, no speeches, no congratulatory message in praise of the founders, no reception, no tea with good things to eat. Many of the members who were present at opening day lived to see the opening of fine new buildings that replaced the first lowly quarters. But it is doubtful that many of today's employees know of the humble surroundings that marked the opening days of their credit unions.

The first function of the new credit union was to receive the savings of the members. The next was to use these savings to meet the credit needs of the members. The mechanics of lending their members' money caused directors and manager alike a little apprehension. One of the best examples of the care with which directors assumed this responsibility can be found in the records of the Coady Credit Union. They were not about to make any errors so they experimented with a trial loan. Mike Campbell, the manager, made an application for a loan, the credit committee held a meeting and passed it, the manager took the amount from the funds and immediately repaid it and the transaction was recorded. The purpose of the exercise had been to "see how it worked." Only after they had seen for themselves that the system worked did the directors give their assent to the granting of loans.

The meetings of the directors during the first few years were serious ones. They were pleased to see the rapid growth in membership and the increasing volume of business but they were constantly mindful of their responsibilities. Some of their concerns may seem unimportant now but they were very real while a credit union was meeting the difficulties of the first years. Some of the credit unions imposed a fine of five cents on those members who did not fulfil their commitment to pay a weekly instalment on shares. Applicants for membership had to be sponsored by a director before being accepted at a board meeting. A director who was absent from a meeting without a good reason could expect a severe reprimand. Directors would personally seek out delinquent members who might be slow in repaying a loan.

The defaulters were few. That could be attributed to the "good sound judgment" of the credit committee members but it was also acknowledged by the directors as a proof of the innate honesty of people. The credit unions thrived because the members were honest in repaying their loans.

The directors were generous in the case of unusual hardship suffered by borrowers. Minutes of Coady Credit Union directors' meetings report the case of a member who died suddenly, leaving his widow and children to live on a pension of ten dollars a month. The poor woman offered two dollars a month from this meagre pension in payment of her husband's outstanding loan. There was no loan insurance in those days. The directors considered the plight of this family and decided to "forgive" the debt. It is significant that they did not consider this a bad loan. There was no intention on the part of this almost destitute woman to default on the debt.

Applicants for loans came fast and early records show how quickly the credit unions began to fill the community needs for credit. Loans were made for the payment of taxes, for repairing homes, for paying doctor bills, to outfit children for school and for many other provident purposes. They were made for productive purposes too. The first loan made by Coady Credit Union was for the purchase of a cow, surely as provident and productive a purpose as was ever put before a credit committee.

Young couples borrowed from the credit union to pay the expenses of getting married and setting up a home. As the president of one credit union put it in his annual report: "Young people have borrowed to equip themselves for an extended voyage on the seas of matrimony."

A frequent lesson learned in the study clubs was the use of the credit union to avoid the prevailing high cost of instalment buying. This form of buying seemed to be the only way in which families could pay for such necessities as clothing and furniture. Few people had tried to calculate just how much interest they were paying under the instalment plan. The sellers were very careful not to disclose the rate of interest. They were very much opposed to a proposed piece of legislation that would require them to state the rate of interest charged every time an item was sold on instalment.

The valuable lesson learned by credit unionists was that they could borrow from the credit union in order to pay cash at the stores, thus in one transaction saving on interest rates and supporting their own institution. Members who had bills in several places were advised to consolidate them, borrow from the credit union, pay all the bills and repay the loan in one transaction. It is interesting to survivors of these early credit union days to find a government agency counselling debtors in trouble and assisting them to manage their finances. Credit unions gave the same service fifty years ago at no expense to the public purse.

With the credit union safely established the directors learned to deal with problems as they arose. The chief of these was the need to educate new members. News travelled fast in a community when a new kind of bank began to do business. Especially in the urban areas new members came forward quickly and the directors felt an obligation to carry on some membership education.

The credit union was a new kind of business that gave each member a vote in the democratic process that was its distinguishing mark. It needed members who understood and appreciated this special role, who would attend the annual meetings and take an active part in policy making and the election of directors and committee members. The early directors were not satisfied that the growing membership fitted this description. For this reason the directors never stopped promoting study clubs. Some preceded the monthly directors' meeting with an hour of study.

The fast coming of the Nova Scotia Credit Union League demonstrates the effectiveness of the study club work that preceded the founding of the credit unions. It was only one year after the organization of the first credit union that delegates were being sent to a convention to discuss the possibility of forming a league. It was so early in the credit union development that it took some ingenuity on the part of directors to find the money to send delegates to the meeting. And at the organization meeting of the League, A.B.

MacDonald held up a package of cigarettes and said, "This is what it will cost annually per capita to support the League." A package of cigarettes cost ten cents, but a credit union that could not afford to pay its manager and which had to put on card parties to pay for office supplies would find it hard to pay the per capita to the League.

At this time of golden jubilee celebrations the work of credit unions is being evaluated in millions of dollars. It should be a time to make another evaluation that comes with no price tag — the hard work, courage and devotion of the pioneers who, in their humble beginnings, had implicit faith in a great future for the credit unionists.

In a speech entitled "Co-operation Builds Men," Dr. Coady in 1947 said in part:

> "One of the primary effects of the credit union is to make people honest. The common people have given loans to the extent of millions of dollars and have not lost anything by them. How can this be explained? It is not easy. It may be group sanction. Men do not want to be guilty of a dishonest act when it violates the delicate and tender bond binding them to their fellow co-operators in a community where the principle of Christian brotherhood dominates the economic set-up."

Chapter VIII
A NEW ROCHDALE OR
"TURNING MERCHANT"

There was a period in the early thirties when the men and women in the study clubs were totally absorbed in the history and the fundamental principles of the co-operative movement. While they were planning to start their credit unions they looked forward to the day when they would do business in their co-operative stores.

The credit union was a co-operative bank with ownership firmly rooted in the membership. It followed naturally that other co-operative enterprises, the retail store in particular, should appear desirable and attainable.

The study clubs very early had become interested in the British social reformers who had attempted to improve the conditions of the men, women and children who worked in the mines and factories at the beginning of the Industrial Revolution. The study club members found much to admire in the work of Owens and King who, in spite of their noble attempts to bring about needed reforms in the troubled society of their time, had failed to establish lasting co-operatives.

The clubs were greatly interested in the causes of the failure of the reformers whose high hopes had not been realized. Their story served as an introduction to the co-operators who *had* succeeded, the twenty-eight poor weavers of Rochdale who, in a pitiful little store in Toad Lane, had laid the foundation for a world-wide co-op movement.

The Extension Department had a wide selection of books on the co-operative movement in England and the study club members read them all. There were two favourites. One, written in the form of a novel, was *Up From the Shadows*, the story of the Rochdale Pioneers by Michael Becker. The author had introduced a love story that was not backed up by history, but that gave an added interest to the work. The other was the well-documented *History of the Rochdale Pioneers* by George Holyoake. This was the classic history of the Pioneers. It was long and detailed but full of humour, vivid descriptions and delightful analyses of the principal characters. It is impossible to over-estimate the influence of Holyoake's work on the early co-operators in Nova Scotia.

The industrial workers, in particular, were interested in Holyoake's work because they found in the story of the Rochdale pioneers something that could be applied to their own conditions. The Pioneers had been desperately poor. Holyoake wrote: "In 1843 a few poor weavers, out of employ and out of food and quite out of heart with the social state met together to discuss what they could do to better their industrial condition. They would commence the battle of life on their own account. They would turn merchants and manufacturers."

The industrial workers in Nova Scotia were probably as poor for their time as the Rochdale Pioneers were for theirs. They too were meeting to discuss what they could do to better themselves. Many of them were "out of employ" and if they were not quite out of food, they certainly were out of heart with the social state. The 1932 report of the Royal Commission on Coal had documented their poverty.

The study club members were intrigued by the way in which Holyoake had summed up the efforts of the Pioneers. The poor weavers had "turned merchants." With their credit unions the study club members had "turned bankers." What was to stop them from "turning merchants" as the Rochdale Pioneers had done?

Holyoake had recorded how the weavers had struggled to collect twopence and later threepence a week, which he said these poor men "did not quite know how to pay." "At length," said Holyoake, "the sum of twenty-eight pounds was accumulated and with this capital the new world which was to be was commenced."

The aspiring Nova Scotia co-operators could easily form a mental picture of the collectors gathering two pence a week from the weavers who were about to commence the new world which was to be. They had their own experience with which to compare the meagre savings of the Pioneers. They had saved five cents and ten cents a week to usher in a new era in finance. If the poor weavers of Rochdale could start a world co-operative movement with twenty-eight pounds, only cowards would shrink from the task of building consumer co-operatives in the Maritimes.

The study club members were impressed by Holyoake's account of the long serious discussions that had preceded the opening of the store in Toad Lane. The weavers had little formal education but they were thinkers and students of the economic and social conditions of their time. Their Sunday afternoon discussion circle in the Chartist reading room, as described by Holyoake, was much like the study club meetings of the St. F.X. movement.

The reasons for the success of the Rochdale weavers, in contrast to the failure of the early co-operators, were well set forth in Holyoake's book and in other books and pamphlets supplied to the clubs. The answer lay in the famous principles of co-operation which had enabled the Toad Lane store to grow into an impressive system of merchandising.

The evidence of history was clear. When co-operatives had faithfully followed the Rochdale principles they had succeeded. When the principles had been violated, the co-ops had failed. The founders of the early co-ops in the Antigonish movement understood very well the importance of the strict observance of the Rochdale rules. The principles were the subject of more study club discussions and more talks by field workers than any other single topic and they were written into the articles of association of each new co-operative. And after the organization of a new co-op the teaching of the Rochdale principles continued, in the study clubs, in the membership meetings, in information to new members and in exhortations to members whose loyalty was seen to waver.

The founders of the early co-ops anticipated the difficulties they would encounter. This awareness had something to do with the decision to start buying clubs. Although the buying clubs were organized out of economic necessity, to help stretch the family food budget, eager co-operators looked on them as a preliminary step toward a co-op store.

The buying club was easy to operate. It simply consisted of the pooling of purchases in order to get wholesale prices, thus by-passing the retail store. It cost almost nothing to run a buying club. One of the members would donate a room in his house, or a barn or shed to store the "orders" until each member came to claim his share. Another member would volunteer to keep the accounts and do the buying. The number of items that could be distributed through the buying club was limited and consisted chiefly of staples such as flour, sugar, etc., and in the rural areas, fertilizers and feeds.

The new co-operators were helped by their membership in the buying clubs, and in the credit unions through which they had obtained experience in working as a group, in conducting business meetings, in handling funds and in communicating with new members in the credit union as the business expanded. The operation of the buying club was so simple that it did not require much attention to the Rochdale principles, but the club did prepare the way for the co-op store because it demonstrated that, by group action, the members could eliminate one source of profit in the long chain of profits they paid to the middleman.

The middleman! He became the chief topic of conversation in the study club as members read and talked about the people who extracted profit from every purchase made by the consumer. In the rural areas one type of middleman was not hard to find. The drover came around regularly to buy cattle. The man who bought fish at one price and sold it at another was also easily identified.

In the towns the middleman was not so plainly visible. Consumers had not been accustomed to give the name of middleman to the merchant from whom they bought their groceries, clothing and other necessities. A woman confided to me that in the first winter during which she took part in the study clubs the middleman was a mystery to her. She thought of him, not in the abstract, but as one real-life person who strode throughout the province gathering profits. He should have been easy to catch. Why was he still on the loose?

There was a period when the clubs studied the nature of profit. The members were impressed with the idea that, in the co-op, profit would be replaced by service and mutual benefit among members. They were beginning to see what they could do to help correct the great default. And so, with a strong belief in the value of the Rochdale principles, they set about the task of forming co-operative stores.

They found that to set up a store was much more difficult than organizing a credit union. The charter members had to collect enough capital to rent premises, stock the shelves and pay wages. The first credit union managers were volunteers who attended to the business on weekends, but the store manager and his clerks had to make their living by minding the store. The credit union manager could learn on the job as the business grew but the man who ran the store had to have a minimum degree of experience. It is remarkable that the new stores were always able to find somebody with enough faith in co-operation to undertake the work of setting up and running the co-op. It is true, however, that the search for managers was made easier by the widespread unemployment.

The first hurdle was to find enough capital. Because the stores started business with insufficient capital they could offer only limited and often inferior service. They could not thrive on the volume of business from the charter members alone. Many new members had to be recruited to provide the volume sufficient to enable the society to provide suitable store facilities and a wide range of goods and services.

The founding members were challenged by the task of educating new members and this meant that they had to find ways to teach the Rochdale principles of co-operation.

The Rochdale principles, as they were studied in the clubs and taught at meetings and conferences, could be found explained in various ways by a number of writers, but no matter how they were expressed, they meant the same. Some writers gave the principles titles, such as democracy, universality, mutuality, equity. Usually the clubs studied them under two headings: fundamental principles and business practices.

The principle of universality or open membership was accepted with enthusiasm, particularly in the urban areas where the constitutions of the labour unions specified that there would be no discrimination in membership.

The principle of democracy or "one member, one vote" was heartily approved. The co-operators were impressed with its significance. To them it spelled economic democracy. We had one lively co-operator who, at every opportunity, demonstrated the importance of the democratic control inherent in co-operative business. He would pretend that he owned one share of stock in the C.P.R. and describe "himself" going to cast his "puny little vote" at the annual meeting of the company.

First he would get a loan from the credit union, although the credit committee would have grave doubts that the excursion to Montreal constituted either a productive or provident purpose. He would put together a passable outfit by borrowing various articles of clothing from the neighbours. He would fortify himself against hunger with a lunch made up by his wife, and he would sit up on the coach to Montreal for the two nights required for the journey at that time. Finally, red-eyed from the loss of sleep and faint with hunger, he would arrive on foot at the palatial hotel which was the site of the meeting, only to find himself sitting next to a millionaire with a few hundred shares and a sheaf of proxies. He would leave it to his audience to judge if his vote would have any effect, for good or ill, on the policies of the C.P.R.

"But," our fiery co-operator would continue, "I could sit next to a millionaire in a co-operative annual meeting and be as important as he. Just thinking about it makes me feel like a millionaire."

In the first few years there was no concern about the Rochdale rule to pay limited interest on share capital. The Rochdale Pioneers had made sure that the value of shares would not go up with the increase in earnings. They laid down the rule with the rate paid on co-op shares should not be higher than the minimum prevailing rate of interest. The rule on limited interest did not cause any problems to the new co-operators in the Maritimes. Most of them were not accustomed to getting interest on anything. Their problem was to gather up enough cash to buy the required number of shares in the co-op. Some of them had to do this by instalments. Some had saved money in the credit union to be ready to buy shares upon the opening of a co-op store.

It was not likely that the early co-ops would deviate from the rule of limited interest because they were not prosperous enough to send shares soaring in value. The founders did not have an interest seeker among them.

It was the principle of equity or of patronage rebates that caused controversy. The founding co-operators accepted this as the most important principle of all — the revolutionary principle that would change merchandising from a system of profit to one of service. It was the magic formula that had brought success to the weavers of Rochdale.

Simply stated, the principle of equity went like this: What is left over, after all the costs of operating the business are paid, belongs to the members who, by their patronage, have provided this fund. The Rochdale Pioneers had decided how this should be done. They ruled that the rebate should be on the basis of patronage. This surely merited the name of equity. Those who made the most purchases at the co-op and thus created the most towards the earnings should get the highest return. Nothing could be fairer.

The theory was excellent and nobody could find any fault with it. In practice, however, this proved to be the thorniest problem with which co-op directors had to deal. Patronage rebates would be available only if a store charged something over and above the costs of operation. The Rochdale weavers had devised strict business rules to ensure that the principle of patronage rebates would work. Goods would be sold for cash only and at current prices.

History had shown that co-operatives that did not follow the rule to sell at current prices did not survive. It was absolutely necessary that the overcharge be made. The threat of a price war hung over new co-ops and the directors knew very well that they had better not fire the first shot. They knew that if they tried to undersell the local merchants, retaliation would be swift and devastating. These well-established businesses, with ample resources, would surely win the war and lay the co-op quickly to rest.

Even with strict attention to the rule the co-ops sometimes had to face serious price competition. For example, this happened to the New Waterford Co-operative in its first year. One of the local stores advertised tempting low prices, particularly in the meat department, with choice cuts selling at what the co-op manager knew to be below cost. The directors called a well-publicized meeting on the Sunday afternoon after the Saturday bonanza of low prices. Speakers were rallied for the occasion. I remember coming from New Glasgow, an eight hour overnight train trip, to be present at this meeting.

The warning was clearly sounded. If the members did not remain loyal to the co-op, it would surely fold up and the attractive deals offered by the co-op's competitor would end. Prices would go up higher than ever to cover the cost of the alluring loss leaders. The enthusiastic meeting and the hard work of the directors saved the New Waterford Co-operative from failing under the tough competition.

When new members were drawn into the fold it was not easy to keep them loyal, in view of the many enticements that drew them elsewhere. Shoppers were accustomed to go from store to store looking for the lowest prices and for bargains. They now had to be convinced that buying at the co-op was the best bargain of all.

New members were attracted first by the promise of patronage dividends. Trouble arose because of their impatience. They wanted to get the rebates quickly. Wise old Holyoake had put it this way: "They do not believe in the end of the quarter" and he had put it in italics. Even in the time of the Pioneers there were members who wanted their "dividends" without any delay.

Complaints came when the patronage rebate was not there at the end of the quarter or came in too small amounts. The new co-ops, struggling for existence, often with insufficient capital and a small volume of business, were not always able to produce the patronage rebate at the end of the quarter. Directors and faithful members had to keep reminding the impatient members thus: "You have been dealing with merchants all your life without ever getting a rebate. Why can't you wait a little longer to get a rebate from your own store?"

The ardent co-operators berated the dividend chaser. They were fond of the story about the little old lady who, while walking home from the co-op with a jug of molasses, fell in the brook, dashing the jug against the rocks and losing the precious liquid. After picking herself up she said to her companion: "Thank heaven, I'll get the dividend."

In those early years while the co-operatives were being slowly developed it was possible to classify members into three groups according to their attitudes toward the principle of the patronage rebate. In the first group were those who thought the theory was great, but who expected too much from its application to their individual store. They looked on it solely as a way of saving money on their purchases. They expected to get high rebates and they wanted them on time. The co-op was just one way in which they thought they could save money and they pursued every other way they could. They were the bargain chasers. They shopped at the co-op when it suited them, but they shopped at other stores too. These bargain chasers were responsible for the prevailing belief that prices were higher at the co-op.

These faithless members provided us with a lot of busy work. We repeatedly had to refute the argument that prices were higher at the co-op. We did this by producing the results of time-consuming price comparison made on tours of various stores. We worked hard to convince the wayward members that comparing prices should be done in a fair way — by considering typical weekly orders instead of specials alone, and by comparing quality, grades and other factors.

The co-ops lost business because of the impatience of those members who had joined with expectations of great savings. They were hard to reach with an educational program. They were not in the study clubs and they did not appear at membership meetings.

The second group consisted of the founders and the firm believers in co-operatives. They completely supported the Rochdale principle of patronage rebates which they accepted as the only way to change the motive for profit into the motive for service, the way which would ultimately give consumers ownership and control of merchandising. They believed that the practice of selling goods at current prices was absolutely necessary to help build up reserves to provide funds for co-operative education and to support the whole co-operative movement as it made progress throughout the Maritimes. These co-operators were not worried because the new co-ops could not afford to pay patronage dividends in the first years of operation. They were prepared to wait. They said that the principle was right and that nothing but good could come from it.

These members left what rebates were declared in the society to add to their share capital. Since they had never before drawn any patronage rebates from the stores where they had made their purchases in the past, they were happy to see the co-op earnings working to build up the general movement.

The third group did not believe in patronage rebates at all. Cost plus was what they wanted. They joined the co-op but never admitted that the patronage rebate was a good thing. Strangely enough, these members remained loyal to the co-op in spite of their objection to the rule to sell at current prices and rebate the overcharge. This group was not very large, but its members talked a lot.

However, among the large group of co-operative officials, directors and educators there was agreement that the Rochdale principle of patronage rebates was essential to the proper conduct of co-op business. Shortly before the opening of the first direct charge co-op in Canada, an informal questionnaire was sent to co-op leaders asking for an opinion on the wisdom of opening co-ops that would sell at cost plus. Those polled included managers and directors of stores and wholesales, provincial co-op unions, agricultural representative, field workers, and others interested in co-operatives.

The responses were overwhelmingly in the negative. Those who replied gave the classic arguments against selling at cost plus. A genuine fear of competition from the private trade was expressed. Those who were concerned with the ultimate goals of co-operation were emphatic in their opinion that the overcharge was necessary to provide for expansion into all the processes of production and distribution of goods.

In recent years some leaders in the co-operative movement have been critical of the Rochdale principle of patronage rebates, some going so far as to say that the movement has been held back because the principle has been followed too closely. The early co-operators found it necessary for the survival of their stores. How the over-charge was used, either as a rebate to the individual member or for the common good did not detract from the wisdom of the men of Rochdale.

The direct charge store has since become widely accepted as a progressive step in co-operative development. But what is possible and suitable in 1985 was not possible in 1935. It was a little frightening to "turn merchant." The co-operators were sustained in their efforts by their faith in the principles that they had been successful for almost a hundred years. There would be no direct charge stores today if the pioneers had not built their co-operatives on the tested rules of the store in Toad Lane.

The new co-operators in the Maritimes found in Holyoake's work not only the story of the Pioneers' success but also the details of the difficulties the weavers had to overcome. These were formidable. They had no legislation to back up their efforts. They could not incorporate and they had therefore limited liability. They could not own land, nor could they get bonding for their cashiers. They could not use their earnings for educational purposes.

The early Maritime co-operators suffered no such drawbacks, but there was evidence that not all the problems had disappeared in almost a century of co-operative history. Inadequate capital, problems in establishing cash trade, severe competition from the private profit trade — all these problems had to be met. Holyoake had said that it took four years before the Pioneers saw their "co-operative wagon toiling surely, if slowly, up the hill."

Our co-operators, comparing their venture with that of the Pioneers they copied, noted that their co-operative wagon was toiling up a hill not so steep and with fewer hazards.

Chapter IX
GENTLEMEN,
LET'S HAVE SOME DISCUSSION

While the people in their study clubs were learning the history and principles of co-operation and the techniques of group action there was much discussion but little dissension. From the start they were determined to improve their economic conditions through co-operative effort, and they were in agreement that the Rochdale principles constituted the basis for success in the co-operative associations they hoped to build.

It was when the co-operatives began to do business that controversy arose. While adhering to the co-operative principles it was possible for the members to take opposing sides on store practices. The chief cause of the dissension concerned the earnings of the co-operative. Two principles were involved in this debate, — limited interest on share capital and the distribution of patronage rebates according to a member's purchases. Everybody agreed that all this was fair and just but there was disagreement on how much of the earnings should be distributed in these two categories. And so they lined up on two opposing sides.

One group wanted to make the co-operative store as tempting as possible in order to "catch" new members. They said that this could be done by paying as high a rate of interest on share capital as the store could afford while remaining within the limits set by law, and by paying as high a patronage rebate as possible. They foresaw two good results from this procedure. Members would increase their share capital in order to earn interest and new members would be attracted in order to get rebates on purchases.

On the opposing side were those co-operative members who took the longer view. They had a different formula for determining what the store could afford to pay in interest and in rebates to purchasers. They thought that before the store could distribute earnings, it had other more important obligations to meet. Earnings should be used to improve service and store facilities, to maintain adequate reserves, to provide funds for future expansion, to pay for educational programs and to contribute to the development of co-operative wholesaling. The members of this group argued that, as co-operators, they were building something that promised to be an important component of a new economic order, and that paying out most of the earnings in small amounts to individual members was not the way to accomplish this. They went so far as to say that a co-operative store could well adhere to the Rochdale principle of equity by meeting all its necessary obligations with its earnings, even though no member received a cent in patronage rebates.

This was the group that strongly advocated that, when patronage debates were declared, members should apply them to increase their share capital. The members in this group were particularly concerned with the "calibre" of the co-op's members. They strongly opposed the argument that the paying of high patronage rebates would induce non-members to join the co-op. They said that the rebate was not meant to be a trap in which to catch new members. Those consumers who joined the co-op with no other motive but that of getting a rebate were not genuine co-operators and the society was better off without them.

The debate was often purely theoretical since the early co-ops did not have much in the way of earnings to distribute. But it is also true that some co-operative stores paid rebates that they could ill afford to pay and at times when there was an urgent need to put funds to better use.

A similar debate went on in the credit unions. Should earnings be used to pay interest on share capital or to give rebates to borrowers? Would keeping interest on capital as high as possible bring about an increase in savings, or should the main concern of the credit union be for the borrowers, who, by the interest they paid on their loans, provided the income? This was a perennial topic at membership and directors' meetings. Usually the annual meeting agreed on a compromise but the topic continued to be discussed wherever directors met either offically or in casual conversation.

Another topic which caused many heated discussions was advertising by co-operative stores. There was no difference of opinion among the founders, who were impressed by the idea that it would not be necessary to advertise in order to persuade members to buy in their own stores. The study clubs had paid careful attention to what was considered an advantage in co-operative buying. The co-operative consumers would be free from the high cost of retail advertising which they had been accustomed to pay to private profit merchants. This freedom from the costs of advertising would set the co-op apart from the other stores in the community.

But when the stores began to do business some managers throught that they had to advertise to meet the competition from the private profit business. They were alarmed when they lost business because customers, members and non-members, were lured by other stores through advertising gimmicks of all kinds such as specials, premiums and loss leaders.

The managers who were in favour of advertising said that they lost sales from those members and non-members who continued to shop around for the best prices. Those members who were against advertising said that efforts should be made, through education, to persuade non-members who shopped at the co-op to join the society. They said that to use advertising to attract new customers was not consistent with the high purposes of the co-operative way. The overcharge paid by these customers went into the store's earnings. In other words, the store made profit from non-member business and it was not in the spirit of co-operative philosophy and principles to seek to make profit through competitive advertising.

After a lively discussion on non-member business, a co-operator who was a stalwart defender of the no-adversiting policy said: "Why should the co-operative borrow the 'gimmick' of our competitors? We have the best gimmick of all — the patronage rebate."

The question that caused the most dissension was the principle of cash trade. While there was unanimous agreement that the Rochdale Pioneers were right when they made the rule that all sales should be for cash, not all the managers and directors were convinced that it was possible to observe this rule.

The chief forum for the discussion of the cash versus credit question was the store managers' annual conference. This was the one occasion during the year when the managers met and when they could talk freely about the problems of minding the store. It was inevitable that credit trade should take up much of the time of the conference. I remember this well because I took the minutes of many of the early managers' conferences.

The rural co-operatives were troubled by the credit problem because their competitors still operated under the credit system and there was a long tradition of credit buying. The rural managers wanted to run credit-free stores but some of them considered that it was impossible to do away with the credit system to which the members had long been accustomed.

Although all the managers' conferences devoted time to the credit problem it was the conference at South River, Antigonish County, that made it the principal theme. There was first a long discussion on the definition of cash trade. The managers of the strictly cash co-ops had no patience with those who had difficulty in understanding exactly what was meant by cash trade. They called it cash "on the barrel head," — the money handed over at the time the purchase was made. These managers admitted of no qualifying circumstances that made it difficult to distinguish between cash and credit trade. Other managers hesitated over this definition. They were not certain that this was the only valid interpretation of the term "cash trade." These were the managers who had members who thought that they should be allowed to trade to some extent on the capital they had invested in the store. They said this was not credit. If the member had fifty dollars' worth of share capital, that money was his. Why should he not be permitted to trade on it?

One of the results of the confusion was that some stores distinguished between share capital and loan capital which bore no interest and could be used for running up bills. This was to meet the objection that capital used for trading was of no use to the store and should not be entitled to interest. The bookkeeping problems that arose from this practice caused some difficulty.

In this connection the South River conference held a long discussion on the possibility of issuing loan capital certificates which could be redeemed at the store for the payment of bills. At least one store had plans to begin this system and there was a lively exchange on how the McCaskey could be used for this purpose. There are no McCaskeys in evidence today. The stores which had these machines that recorded accounts receivable have long ago consigned them to the dump, basement or attic. On one occasion a manager was heard to say that if all the McCaskeys were abolished, the problem of accounts receivable would disappear, and he wished that some kind of total destruction would befall them.

It was also suggested that these loan certificates would be used as security for credit union loans. At that point the wisest of the managers collectively lost their tempers. They deplored any scheme that was designed to make any form of credit acceptable at the co-op. All this useless talk about loan capital and certificates meant that the store would assume the costs of credit trade, thus invading the field of the credit union whose sole and proper function was to provide credit. The managers were emphatic on this point: No matter what form the delay of payment for purchases took, it violated the principle of cash trade.

The comments were sharp. One manager said that the stores which did not give up credit trade were "headed for the rocks." Another commented that the use of credit was causing the co-ops to abandon the principle that all the members were responsible for financing the store, and that in those stores where credit was allowed the "solid" members were paying the cost for those who ran up bills.

One astute manager said that a change to a cash policy might result in a drop in sales at first, but that the remaining sales would be "sound business." Another said that the co-ops had been put on trial and found guilty of violating the Rochdale principles. An agricultural representative closed this part of the discussion by saying: "We have taken the wrong road by failing to observe the rule of cash trading. We must now turn back and find the right road." One manager said: "I have an agreement with the manager of the credit union. He doesn't sell sugar and I don't give credit."

One manager described how his store had gone about establishing a cash policy after it had fallen into the trap of giving credit. The members at the annual meeting had decided to enforce the rule, beginning on a specified date, and had ordered that three special meetings be held in the month previous to that date. At these meetings directors gave a full explanation of the reasons for the change. The plan had been successful and the once credit-ridden store was on a strictly cash basis and doing well without a McCaskey.

The South River conference devoted time to the role of the credit union in assisting the store to maintain the rule of selling for cash, particularly in the rural areas where members were on a seasonal income. The meeting instructed the resolutions committee to draft a resolution on the subject of cash trade. The resolution presented to the conference deplored the practice of credit trading against capital invested in the store as a real obstacle to the development of the co-operative movement in Nova Scotia and asked the N.S. Co-operative Union to institute a program designed to establish cash trading.

There was little the N.S. Co-operative Union could do to establish a program that would abolish credit trade. Only by education could managers and members be persuaded to abide by the cash rule. The problem continued to be a real threat to the stores and two years later the manager's conference had another debate. The same problems were discussed but there were reports of successful changes from credit to cash. The managers of the wholesales made strong appeals to the store managers. Credit was a serious problem to the wholesale since the members' stores had equity in the wholesales and traded on that investment. J.T. MacDonald, manager of Cape Breton Co-operative Services, emphasized that cash trade could improve the economic position of both the individual co-op and the wholesale and that this improvement was more important than the increased sales that might be obtained from credit trade.

Time took care of this major obstacle to the progress of the co-operative stores. Credit trading in retail stores has long since disappeared from the economic scene. Managers, directors, and members who have never had any experience with credit trading in co-ops may find it difficult to understand why there was so much commotion and confusion over this Rochdale rule. But credit was the co-operatives' most serious problem. It was a constant source of worry to directors and managers. It brought co-op stores to the brink of disaster. There is no doubt that it greatly hindered the development of sound co-operative business.

Another topic which caused much discussion, with the least dissent, however, was the relative roles of producer and consumer in the economy. It was inevitable that in the work of the study clubs each vocational group would examine the economic conditions of the other groups as well as its own. This led to the question: "Who is more important in rebuilding the economic order, the producer of goods or the consumer who, in the final economic transaction, pays for the total cost of production, transportation and merchandising?"

On the highest level, what some called the 'philosophical plane" the primacy of the consumer won the debate. But the study club members talked about producer and consumer co-operation in a more practical way. This was because of that common mischief maker, the middleman, who had traditionally stood between the primary producer and the buyer of his product, whether in the first transaction at the farm or in the final one when the consumer put down his money on the store counter.

In the study clubs there was solid agreement on first principles. The primary producer was entitled to a return sufficient to pay the costs of production and to enable him to support his family on a reasonable and comfortable standard. On the other hand, the standard of living of the industrial worker depended only partly on his wage and to a great extent on the prices he had to pay for consumer goods.

In an ideal situation fair prices to the farmer and fisherman should prevail along with fair prices paid by the industrial worker. But the situation was far from ideal. The price spread from the producer to the consumer was a serious impediment to the attainment of a good standard of living for the industrial worker who had to buy everything needed to keep a home in operation.

The industrial workers learned in their study clubs that the only way they had to reduce the price spread was through their consumer co-operatives. They expected their farm cousins to do the same. Genuine co-operators were disappointed to find that some primary producers who, while they sought to raise their standard of living by co-operative group action, continued to support the private profit system in which the middleman prospered. These "half-co-operators" were the loafers in Dr. Coady's "vestibule of co-operation."

All these questions, — interest rates, patronage rebates, credit trading, advertising, price spreads, the middleman, producer-consumer relations — added hours to the length of meetings and provided discussion that continued among small groups of people who were not in a hurry to get home after a meeting.

The longest meeting on record lasted all night. I was not there, but I got an hour-by-hour description from A.S. MacIntyre who *was* there. There was spirited objection to a resolution by a group of men who were determined to stop its adoption. The meeting heard speech after speech by the dissidents, abetted by the chairman who was on their side and not entirely impartial in his conduct of the proceedings. At one o'clock in the morning one of the agricultural representatives, who was noted for his long-windedness, arose to speak. He went on for an hour and a half and then sat down with the invitation: "Now, gentlemen, let's have some discussion." As the discussion had been going on since eight o'clock the previous evening, the call to speak up was hardly necessary. The talking continued throughout the rest of the night. The train departed from Sydney at six a.m. and the gentlemen from Antigonish, Mabou and Halifax were just in time to get on board, without breakfast.

Chapter X
THE BIG PICTURE

As the number of co-operative stores increased the co-operators began to talk with enthusiasm about the "Big Picture." It became the slogan wherever co-operators met and where the ultimate goals of the co-operative movement were discussed.

In the big picture were all the credit unions and co-operative stores. There were also the larger associations of these basic units — the credit union leagues and the co-operative wholesales. But the big picture was incomplete. The eager co-operators wanted to put in the picture more powerful credit union organizations and that most desired economic institution, the co-operative factory. They wanted a larger share of ownership in the nation's business.

The idea of the big picture was not new. At the beginning of the movement Dr. Coady had taught that the credit unions and the co-operative stores were only the first steps in the program that would enable the people to change the economic system. The credit union was a good thing, but by itself it would not bring about a new era in finance. The small co-operative store was more important as a basis for consumer control of merchandising than for the small savings it offered to its members. The power of the local store lay in its union with other stores to enter into wholesaling and manufacturing.

There was a full circle in the way the members of the early study clubs perceived their role in the co-operative movement. They were dreamers as were the Rochdale Pioneers whose example they chose to follow. The men of Rochdale had great plans that extended far beyond the limits of the Toad Lane store. "As soon as possible," they said, "this society shall proceed to arrange the powers of production, distribution and government." Holyoake wryly remarked about this plan: "No nation had attempted it and no enthusiastic has carried it out." But although the Rochdale Pioneers never achieved the goal of arranging government, they did make significant progress in distribution not only in England but throughout the world. Not all the co-operators in the Extension movement predicted that they would "arrange" government, but they all hoped to change production and distribution and thus gain some control of the economy.

The pursuit of the dream began with the hard work of building the primary structures on which the more complex institutions would depend, and for a time the co-operators were fully occupied with setting up local credit unions, buying clubs and co-operative stores. As the preliminary work progressed the co-operators found their dream again, and once more the study clubs were making grand plans. This happened rather quickly. The first credit union was organized in 1933 and only one year later plans were being made for a credit union league. By 1938 the Cape Breton Co-operative Services and Maritime Co-operative Services were drawn into the picture.

It was then that the educational program entered a new phase with the emphasis on the importance of the local co-ops as partners in the wholesales, gathering strength to enter into manufacturing. The Big Picture was a topic at all membership meetings, conferences, short courses and wherever co-operators met in informal gatherings. Co-operative directors and the members who had come through the study clubs were excited about the vision they saw in the Big Picture. They then had to convince co-operative members that they had power in the simple transactions they made over the store counter.

New co-operative store members, who had not had the benefit of study in the clubs, understood that in the co-operative they could obtain as a patronage rebate what they had previously paid as profit to the co-op's competition. They then had to learn that the method by which this rebate was obtained was more important than the rebate itself.

Just as a member could get a patronage refund from his co-operative store so could the store obtain a rebate from the co-operative wholesale. Shopping at the co-op achieved a new importance when the wholesale was put in the picture. The members' purchases would build up the volume of business done with the wholesale and thus bring closer the expansion of the movement into manufacturing. The day of universal rejoicing when the co-operative factory would be put in the big picture was hastened by every single purchase made at the local co-operative. Then the Big Picture would be complete with co-operatives "covering the Maritimes and the wheels of industry turning in our own factories," as predicted by Dr. Coady.

Then the problem of loyalty arose once more. Just as it had been necessary to exhort and persuade members to be loyal to the stores, it was necessary to persuade managers of co-operatives to be loyal to the wholesale. Experience soon showed that managers looked for "deals" with the same zeal that shoppers looked for bargains and consequently, co-operative members were often disappointed when they looked for co-op label products on the shelves in their stores.

Despite early difficulties co-operative wholesaling made progress. Co-operative leaders hoped that it could help to reconcile differences between producers and consumers. Those who had taken part in discussions on the "philosophical plane" of producer-consumer relations now had something practical to discuss.

The wholesale bought what it could from co-operative producers and sold to co-operative consumers. Efforts were made to increase the kinds of products that could be called co-operative "all the way." Stores in town and country were joint owners of the wholesale and for the first time farmers, fishermen and industrial workers owned something *together*.

The unifying effect of the co-operative wholesale depended to a great extent on the producers' loyalty to the consumer co-operative. Producers had to believe that their support of the co-op store was as important as the gain they could make from co-operative marketing. Dr. Coady compared producer and consumer co-operation to the arms of a nutcracker and the astute co-operators had no difficulty in identifying the nut it was supposed to crack.

Credit union and co-operative conferences brought the different voational groups together. An incident that happened during the coalminers' strike in 1947 showed how the understanding between the vocational groups had grown.

It was the first coal miners' strike in twenty years. In spite of the loss of a large number of the young miners to the armed forces, the men had established records in production of the coal needed for the war effort. During the whole of the war they had worked for the wage rates of the depression years. When at the end of the war they looked for a contract with an increase in wages they were turned down and a strike followed. Public sympathy was with the miners and support was generous.

The industry was completely shut down. No coal was mined and none was moved except to hospitals and schools. Pickets manned the highways from the mining towns to keep bootleg trucks off the road.

It was early spring and Cape Breton Co-operative Services had bargained with the coal company for the purchase of stone dust used in the pits as a protection against explosions. The stone dust contained a high percentage of lime which farmers needed for the land and they depended on the wholesale to fill their orders. Since absolutely nothing was moving from the company property it seemed that the farmers would not get their limestone.

Ted MacDonald, manager of Cape Breton Co-operative Services, decided to try to get his orders filled. He referred the matter to the St. F.X. Extension Department which supported him in an appeal to the United Mine Workers to permit the movement of the stone dust. He came to union headquarters in Glace Bay to meet with the union officers and representatives of the men on the picket lines. Twenty men were present at the meeting and the decision was unanimous to grant the wholesale's request.

These men were members of credit unions, some of them directors and veterans of the study clubs. They had met farmers as friends at credit union conventions and co-op rallies. The pickets were ordered to let the wholesale trucks pass through. Nothing else moved from the mines while the strike lasted.

As the business of the wholesales increased so did the plans for processing and manufacturing. Co-operative members were continually reminded that their loyalty would help to put co-operative factories in the big picture. It was a beautiful dream! In the ideal co-operative factory there would be perfect harmony between producer and consumer. Joint ownership would be vested in the producer and the consumer through the retail stores and the wholesales. For example in a food processing plant, patronage rebates would be distributed in three ways: one-third to the consumer, one-third to the producer and one-third to the plant. A completely co-operative product would be on the store shelves. The ideal organization had been achieved in other parts of the world, notably in Waukegan, U.S.A. Why not in the Maritimes?

The topic of co-operative factories inevitably introduced into a meeting the history of Nova Scotia's poor record in maintaining manufacturing industries. The co-operators were accustomed to hearing the sorry news that an industry had either collapsed or moved to Ontario.

One of the reasons traditionally given for the failure of Maritime industries was that the Maritime market was too small to absorb the products of the factories. The co-operators did not believe that. They believed, as Dr. Coady had taught, that the Maritimes should be considered as an economic unit which could support many kinds of industry.

They longed to draw a flour mill in the big picture. Dr. Coady often talked about a flour mill, situated at the mouth of the St. John River, owned by co-operative producers and consumers and distributing flour to hundreds of co-operatives throughout the Maritimes. It would have been difficult to convince these co-operators that Maritime people did not eat enough bread to make the dream come true.

And if the Maritimers could not consume all the products of a given industry why could they not export them? The second reason usually given for the flight of industry to central Canada was the high cost of transportation between the regions. The co-operators objected to this argument because, by their calculation, it was just as far to go from Toronto to Sydney as to make the trip in reverse. As one of the oldtimers put it: "It would take just as long to walk between the two places, whether coming or going." Transportation costs should be the same on goods going west as on goods coming east.

At many a meeting this question was asked: Why, poor as we are, are we expected to pay the transportation costs on the goods we buy from central Canada, while the consumers in that region can't import from us because of high transportation costs? Why are our raw materials shipped out of the province to be manufactured in other parts of Canada and shipped back as manufactured goods?

There was the case of the travelling cucumbers, for example. An Ontario firm commissioned Maritime farmers to grow large quantities of cucumbers. At harvest time the cucumbers were shipped to Ontario where they were made into pickles. After the time required for going back and forth those cucumbers ended up in pickle jars in the co-op store at prices that covered the cost of the journey. The co-operators thought Maritimers could surely make their own pickles, thereby adding a processing plant to the big picture.

The forests were often quoted as the source of raw material shipped from the Maritimes and returned in the form of manufactured goods. Dr. Coady would say: "the maples are sighing and pleading with us to cut them down and make beautiful furniture out of them. But we don't listen to their sighs. We prefer to import spindly little chairs that break when a big fellow like me sits down."

The Big Picture was not completed in the Maritimes. The co-operative factories were not drawn in. The future did not bring the realization of the co-operative dream. Competition arising from government subsidies to central Canada manufacturers have remained a serious obstacle.

The true co-operators in the area never admitted that what they tried to do was impossible. What they tried to make happen could have happened. One thing is certain. The devotion and hard work of the co-operative pioneers in the Maritimes was sustained by their belief that what they saw in the Big Picture *could* become a reality. If they had not believed this, the progress in co-operative development would not have gone beyond the early stages.

Chapter XI
"WOMEN'S WORK"

From the beginning of the Antigonish Movement women took part in the study clubs. They were in the midst of all the intellectual activity generated by this new program of adult education. The clubs met in their kitchens and parlors. When friends and neighbours met conversation often turned to the new learning that had come into the life of the community.

During the period from 1933 to 1944 the St. F.X. Extension Department directed a special educational program to women. The program was known as "Women's Work" and it involved the formation of clubs in which women studied subjects that were of special interest to them. By 1938 there were three hundred and fifty study clubs composed entirely of women. At the same time there were "mixed" clubs of men and women and some women belonged to two clubs, one of each kind.

The subject matter studied in the women's clubs came under two headings. One was related to women's primary function as homemaker, the other to women's role in the new co-operatives that were springing up from the adult education program.

The topics included homemaking, nutrition, handicrafts, better buying, consumer education and, for the rural clubs, self-sufficiency in the production of food. This emphasis came from the self-sufficient rural life orientation that had marked the origin of the Extension Department. The Extension Bulletin, through its editor, George Boyle, continually advocated "back to the land" and handicrafts as the way to a good life.

There was special emphasis on education in homemaking because that was the principal and often the exclusive occupation of women. The emergence of women from the home into the labour force had not yet begun. Those women who did work outside the home fell into a few categories. There were teachers, nurses, stenographers and clerical workers in stores and offices, but, with very few exceptions, these were single women. As late at 1940 a woman employed in the Nova Scotia civil service would be promptly fired when she got married. I have first-hand knowledge of this because it happened to me. In 1940 I was teaching economics in the provincial correspondence study division. When I got married that year I was at once reported to the division by a man who was seeking my job. In a letter expressing high commendation for my work, I was informed that my spy was succeeding me.

Although the homemaker was not "gainfully employed" she had a full-time job. It is doubtful that any of today's "liberated" women could carry on their careers if they had to cope with the conditions that prevailed in the homes of the thirties. These homes operated on woman power. Most of the homes in the industrial areas had electric lighting but for many that was the extent of electric power. There were few washing machines in the farm or town homes. Clothes were washed by hand on the washboard with the aid of the hot water boiler sizzling on the stove. The clothes dryer had not yet come into the homemaker's vocabulary. The women knew that there was a machine that would wash clothing, but they did not dream that there was anything but the sun, the wind and the kitchen stove to dry it. There was no permanent press clothing, and washing was only one of the maintenance chores that included starching, ironing and mending. Homes did not have refrigerators and except during the cold winter months, food could not be stored. There were no "convenience" foods. Homemakers made their own cakes without benefit of mixes.

All they asked for was the necessary raw material. Getting the butter and eggs and milk and sugar was the hard part. Once supplied with these things, the women were happy to make a cake.

It is probably true to say that many homes that are now on social assistance are better equipped than the workingman's home in the thirties.

Many items, especially children's clothes, were made on the home sewing machine. It was the era of the flour bag, bleached at the cost of much time and effort, and made into various articles of clothing. It took a lot of work to produce a pair of bleached embroidered pillowcases from flour bags, as a gift for the bride, but the happy recipient fully understood that it was the product of a labour of love from a friend.

The depression years were particularly difficult for the women in the towns. The thirties brought unemployment for the men but there were no unemployed women. The job in the mine or the plant was gone but the work at the stove, the wash tub and the sewing machine did not go away. For the homemaker there was no such thing as part or idle time.

Farm women were a little better off only because the farm produced food for the table, but there was very little cash and the homemaker had the same struggle to clothe the family. The lucky children were those who had sisters or aunts in "Boston" from where there came boxes of clothing, cast-offs from wealthy employers.

From these impoverished homes came the women who joined the study clubs. The most vivid memory I have of the many study clubs which I attended is the cheerfulness of these women. They seldom complained. When they talked about the difficulties of their daily lives as homemakers, it was not because they were sorry for themselves but to help one another in the attempts to make life better for all.

The feminists of a later time might say that these women were too passive and too attached to their home life. But they did what they had to do and it was right for their time. The family came first. The children were fed and clothed as well as the income permitted. The women of the thirties were the heroines of a desperately hard age. That they managed to keep their homes resonably secure against the evils of unemployment, low income, illness and inadequate medical care is to their everlasting credit.

It was to these women that the St. F.X. Extension Department directed its program on "women's work." It was headed by two indefatigable, enterprising women, both home economists and Sisters of the Order of St. Martha. Sister Marie Michael was appointed director of "women's work" in 1933 and she was joined in 1935 by Sister Anselm (Sister Irene Doyle) who was given the special assignment to promote handicrafts. It is amazing that these two women were able to promote, supervise, teach, and organize so many activities in what was called "Women's Work," and at the same time perform a variety of tasks in the general program of the Extension Department.

Besides directing the women's program, which until 1934 included taking charge of the women's section of the Extension *Bulletin*, Sister Marie Michael gradually assumed complete charge of the Extension library, a job which required a large volume of correspondence with borrowers all over the study club area. She was one of the main resources for the time-consuming visitors who came from many parts of the world. She was often called upon to speak at meetings, rallies and conferences.

Sister Anselm was also so versatile that she could not escape being recruited for new jobs as they arose. From 1936 to 1938 she did a great deal of work for the Nova Scotia Credit Union League. She attended to the correspondence from credit union managers who had questions about procedures and who needed supplies; she kept the League accounts and produced the monthly financial report. Her field trips had to be arranged to permit her return to the League office on time to prepare the reports. When the credit unions were being organized in Prince Edward Island and New Brunswick she assembled and packed bookkeeping sets and made them ready for shipping. On almost any visit to the Extension office the two Sisters would likely be found surrounded by cartons, wrapping paper and twine and stacks of books either from the library or from the credit union supply cupboard.

Sister Anselm was also artist-in-residence because she was skilled in drawing. What spare time she had must have been all taken up with designing and drawing posters, covers for pamphlets, linoleum cuts and other art work.

Sister Marie Michael and Sister Anselm participated in all the staff conferences, the leadership schools and other important meetings held by the Extension Department. It is certain that if the women's division of the St. F.X. Extension Department were operating today as it was in the thirties, these two women would each have assistants and secretarial staff.

The first task of the women's division was the procuring of study material. From the desks of Sister Marie Michael and Sister Anselm there came an abundance of pamphlets and lessons on all aspects of homemaking: nutrition, home management, buying and handicrafts. Sister Anselm was in charge of the handicrafts program. This included providing and distributing study material, teaching at short courses and conferences, and organizing exhibits.

The Extension Department promoted handicrafts for several reasons. It was thought to be an effective means of arousing interest in study clubs. There appeared to be an advantage in combining practical activity in crafts with reading and discussion in the general educational program. We sometimes had an ulterior motive in inviting a group of women to come to a meeting or a course at which some popular craft, such as glove-making, was to be taught. Having captured the women and initiated them in the craft we would then turn their attention to the study of the credit union or co-operative store. Sister Anselm recalls that on one such occasion the wife of the community's chief merchant found her way to the meeting.

It was also considered worthwhile to encourage women in the creative effort of doing a piece of artistic handicraft. The women were encouraged to produce handicrafts for sale as a means of increasing the family income. To assist in this project the Extension Department promoted exhibits, first on the local level and later at the Rural and Industrial Conference. Each exhibit was a means of education for women interested in craft work and on occasion for exchanging ideas with others who had mutual interests.

The handicraft program required field work. Sister Anselm went forth for short courses sometimes for a day or two, sometimes for a week and occasionally for a whole month in a community. The program discovered women who were skilled in various forms of craft work and who were pleased to teach their skills to others.

In the longer courses women were able to learn useful sewing skills. Some of the projects were ambitious. Some women brought to the classes articles of clothing which they wanted to repair or make over. One woman brought her husband's suit which looked so shabby that Sister Anselm and the class decided to "turn" it. They took the jacket apart and remade it with the inside turned to view, looking much better than the former worn outer side. It was an operation that even an experienced tailor might have hesitated to try, but it was accomplished successfully.

As time went on it became evident that the handicrafts program required more staff and money than the Extension Department could provide, but because it was never intended that it should be abandoned, alternate methods of support were sought. Sister Anselm took careful steps to ensure that this would happen.

In 1942 a two-day handicraft conference, under the sponsorship of the Extension Department, was held at Antigonish. It was attended by sixty-five delegates including handicraft experts from outside the province. From this conference there came a policy committee with a mandate to draw up a statement recommending to the Nova Scotia government the development of arts and crafts in the province. Following the presentation of this recommendation to the Nova Scotia government, a similar conference was held, under the auspices of the Department of Industry in Halifax and the result was the establishment of a handicraft division under that Department.

Thus there came the end of the handicraft program of the Extension Department. The experiment had been successful in spite of the shortness of staff. Hundreds of women had learned to make beautiful articles for their homes without incurring much cost. Others had added to the family income through the sale of their work, and although the plan had not worked perfectly, women had been introduced to the general study program through their interest in learning to do craft work.

During all this time the three hundred-odd study clubs were busily engaged in the study of nutrition. To supplement pamphlets available from the Departments of Health and Agriculture, Sister Anselm and Sister Marie Michael produced study material specially suited to the incomes of the women in the study clubs.

During one winter study season we used an outline for a week's menus devised by Sister Anselm and based on the food budgets of the industrial worker's family. She must have worked very hard on this project. The note attached to a huge package of this material sent to the Glace Bay office read: "It will require a lot of ingenuity on the part of the women on this food budget to make these meals the least bit interesting." The women, undaunted by the challenge, followed the course with great enthusiasm, supplementing the menus when the week's food allowance made this possible. They enlisted the support of the family, posted the day's menus in the kitchen, and explained the rules of nutrition that governed each selection.

Then Sister Anselm prepared the pamphlet: "The Fat of the Land" which integrated homemaking, nutrition, meal planning and the purchasing and growing of the family's food supply. It was particularly addressed to the rural women, suggesting how to use the resources available in the local community and emphasizing the value of natural over processed foods. The pamphlet was full of information for rural women to help them to manage the many duties of the farm wife to the best advantage. But although it had a special appeal to the women in the farming areas, it was used in study clubs all over the industrial areas where it was received with enthusiasm by women who had never done any gardening but who could appreciate the valuable nutrition lessons it contained.

On occasion we ran into a little opposition from men on the speaking circuit. At a meeting in one of the rural communities a man speaking on a agricultural topic cautioned the women in the audience to grow more potatoes and turnips and not to waste time on "fancy stuff." The women in the hall, fortified by their study of the "Fat of the Land" and well aware of the vitamins contained in the speaker's "fancy stuff" proceeded to give him a lesson in nutrition.

A woman whose husband was employed full time, although on a low wage, said that there was an argument over the food bill each week. The husband complained that she was spending too much money on milk for the children. She finally brought hostilities to an end by preparing a breakdown, as complicated as she could make it, of the nutritional value of milk as compared to the other items on the food order. A quart of milk cost eleven cents. She challenged him to suggest how eleven cents could be better spent. There were no more complaints.

One good result of the nutrition program was the appearance of home canned foods on pantry shelves. The information which the women gained in nutrition and good buying was put to good use in this project. When the canning machine became available the Extension Department organized the women for group buying and gave demonstrations on this new canning method. The women knew a lot about canning but the machine was new to them. The beginning of the war brought about added interest in canning, most of which had to be done without sugar, which was rationed.

Some of the demonstrations were done on a large scale. At Dover, Sister Anselm went to perside over a big canning project. The women made a fire on the beach and canned, among other foods, a large quantity of apples which they had obtained free from the Annapolis Valley. The cans were processed in huge pots used by the fishermen for tanning nets. One of the women brought meat which she said was pork, but which everybody knew was moose. She told the story about how she had been reported to the game warden or the R.C.M.P. She was not sure which branch of the law was in charge of investigation, but the man who came to the house was very polite and did not ask any specific questions. He commented on how good her roast (which was in the oven cooking) smelled. "And I said to him," she said, "if you was a piece of pork and you was cooking, you would smell good too."

It was a clever answer. She did not say the roast was pork, but then she did not say it was not! Sister Anselm made no comment on the subject of hunting out of season. Knowing how scarce meat was on Dover tables, she was all in favour of shooting more moose.

The nutrition program brought much important information to hundreds of women. It helped them to make the best possible use of the meagre funds they had for providing family meals. And from the study on nutrition there came a new interest in consumer education.

The women in the study clubs were primarily interested in the relation between good nutrition and the cost of food. With their limited resources they could manage to provide sufficient food for the family but there were food products essential to good nutrition which they could not afford. It followed that the basic foods became the subject of much discussion. The women considered the effect both in the loss of nutritional value and the increase of costs brought about by the processing of some common foods.

The Extension library distributed among its most popular books the works of authors who exposed fraud in the marketplace. Books like Stuart Chase's *Hundred Million Guinea Pigs* had been widely read. These books exposed the ways by which customers were manipulated into buying worthless and even harmful foods at prices that brought huge profits to the manufacturers and earnings for the advertisers.

When the women read these works they could relate the examples to their lessons in nutrition. For example, many women were still clinging to the custom of serving oatmeal porridge, some because the family liked porridge, but others because they could not afford the processed puffed cereals. It could be said that the whole series of lessons came out of the porridge pot. We built a demonstration program around cereals to demonstrate the superior nutritional value of basic cereals compared with the same cereals in processed form. The information which we obtained from wandering around in co-op stores, notebook in hand, was made into lessons and demonstrations. The women were exhorted to beware of processing which changed the form and appearance of food, but which often destroyed essential nutrients and commanded exhorbitant prices.

It was to be accepted that, when women looked for a place to put into effect the lessons they had learned in nutrition and consumer education, they should find that place in the co-op store. But a quick tour of any co-op store showed that it was not different from any other store which the members had left to get a new and better way of buying. The idea that the co-op store should be different took root. Why should the co-op store carry a dozen or more brands of the same product, all composed of the same essential ingredients and differing only in the name stamped on the package? Why should the co-op member pay the increased cost of the handling, the shelf space, the bookkeeping and other costs to the store because of the large variety of brands?

In their clubs the women learned that the co-op should tell the truth about the articles it offered for sale. This meant that consumer education should be part of the service it offered to its members. There were good authorities for bringing consumer education into the store. There was the mandate from the Rochdale Pioneers who had stated firmly that the co-ops should sell only "good goods." In 1844 that was not very difficult. Selling good goods meant making sure that the foods were not adulterated and this could be determined with the naked eye. It meant giving honest weights and measures and that could be accomplished by an honest clerk watching the scales.

Selling good goods ninety years after Rochdale was much harder. The adulteration of food had become a complicated procedure involving all sorts of additives. Exactly what was in the package of processed food? How many of the sixteen ounces registered on the package of rice shot from guns were taken up by air or some other filler? What was in the patent medicine besides alcohol or water, or both? The goods in the co-op store were not "good" if they were not what the advertisers claimed them to be, if they were harmful to health, if they had been stripped of their nutrients and if they were overpriced.

There was another authority for consumer education in the store. Dr. Warbasse, the famous U.S. co-operator, had said in his book *Co-operative Democracy* which had reached the status of a test book in the clubs, "The co-operative store should be a centre of education." The more thoughtful women in the study clubs were sure that managers, staff members and directors would have to give their full support to a program of consumer education if the member was to be served honestly with good goods.

The "Women's Work" came to include as much consumer education as could be fitted in at meetings and conferences. Displays and demonstrations were not confined to women's meetings, but were used at many short courses and conferences. The response was usually good except that from some managers. At a conference where some disturbing facts about the harmful effects of certain patent medicines were quoted from reports of qualified medical authorities, one manager stood up to say: "But there is a good margin on patent medicines."

Emphasis was placed on the co-op label as an important feature in the consumer education program. The Extension Department published a study club outline in a pamphlet entitled "Shopping Basket Economics" which presented women with a vision of their power as consumers. It made a case for the co-op label as a way to improve nutrition by quality food and to save money. Its chief message was that choosing co-op labels became a powerful way to participate in the process of marketing. By the consolidation of purchasing in the co-op stores through the wholesales there would eventually be enough demand for the co-ops to establish manufacturing plants. The consumer education program, so closely related to women's everyday participation in buying transactions, would help to make them aware of their power to complete the Big Picture.

The active period of "Women's Work" came to an end in 1944. The war had brought many changes in the homemaker's life. The bread winners were now employed and there was, for the first time in many years, assurance of a pay envelope at the end of the week. It was a transition period during which women began to move into employment outside the home in industry and service jobs.

The active period of the women's study clubs was over. There were no structures like credit union buildings and co-op stores and assets in dollars by which to evaluate the results of the program of "Women's Work," but its good and lasting effects were not any less important. By providing stimulating study in handicrafts, nutrition and consumer education, and in the development of credit unions and co-operatives, the program did much to sustain women through the bleak depression years.

When the women's division of the Extension Department was closed its faithful workers could believe that, to some extent, progress had been made towards the goal as expressed by Sister Marie Michael at the 1938 Rural and Industrial Conference: "Women must be thoroughly imbued with the vision of a better world and the means to obtain it."

Chapter XII
THE WOMAN WITH THE BASKET

"The Role of Women in the Co-operative Movement" was a topic that appeared frequently on the programs of the many conferences held by the St. F.X. Extension Department.

At the very first Extension meeting in which I took part my assignment was to make a special appeal to women to join in the study club movement. The appeal was made on two premises. The first was that each community needed to put into use all its resources, including the ability and talent of both men and women. The second was that women, as the chief spenders of the family income, should use their purchasing power to bring about desirable reforms in the economic system.

Very early in the development of the new co-operatives it was recognized that their success depended largely on the support of the women in the family unit. The rapid growth of the credit union happened because women were the chief savers. When coal miners deposited in the credit union the sum of twenty-five cents from the week's pay, it was the homemaker who, by some sort of economic miracle, had managed to save that amount from the household budget.

It was easy to get women to support the credit union because its good effects were immediate and easily identified.

As soon as the credit union began to make loans from the small savings of the members it stood out as a friend in need. One has but to study the list of those early loans and the purposes for which they were made — getting the children ready for school, repairing the leak in the roof, paying the bill at the drug store — to understand how the credit union helped to ease the burden of worry that beset the mother of the family. The women who had taken part in the study clubs were the first to use the credit union loan to save the exorbitant cost of installment buying. And many a new bride met the cost of furnishing a modest new home by means of a credit union loan.

In the co-operative store the women were the key to success. Fishermen could sell their lobsters and farmers their lamb through their marketing organizations with nothing more than moral support from their wives and daughters, but their co-op store would fail in its very first year if it did not have the active support of the women of the family.

It was not as easy to convince women that they should be loyal to the co-operative store as it was to create enthusiasm for the credit union. The advantages of membership in the co-op store seemed remote to the homemaker struggling with the food budget. Dealing at the co-op sometimes required a drastic change in buying habits.

There was no problem with the women who were active in study clubs. They were loyal to the co-op because they believed in the principles and philosophy of co-operation. But new members, who were attracted to the co-op because of possible savings in food bills, were not as much interested in social dividends as they were in saving money on the week's grocery bill. It was from these women that complaints came about high prices and low rebates. The women were the "higglers" described by Holyoake. In the Maritimes they were known as dividend chasers. The men would rightly blame a co-op's slow growth or its decline on the women who shopped from store to store looking for premiums, loss leaders and specials.

There was also some resistance to the cash rule which the co-ops tried to enforce. This was particularly true in the rural areas where income was irregular and seasonal. Store members had to free themselves from the pattern of store credit with the merchant who had "carried" them in the past. This is why the credit union was an absolute requisite to the success of the co-operative store. We had a slogan in the women's study clubs: "You don't buy goods at the credit union and you should not buy credit at the store."

In the study clubs, the meetings and the conferences, the power of the "woman with the basket" was emphasized. No speaker, man or woman, lost an opportunity to make reference to the economic power of women because they were the nation's spenders of the family income. It was no revelation to women to be told that they were the country's chief buyers. They were conscious of this fact as they struggled to run a home on what income they had. But it was a revelation to be made aware of how powerful their spending was if done in a co-operative way.

This faith in the power of the consumer had been held by social reformers for the last hundred years. William King, the famous English co-operator, had said: "Your purchasing power is your greatest power, provided it is organized. Unorganized, it is a way to keep you in subjection."

The field workers in the Antigonish Movement could say it no better. Only Dr. Coady had improved on King's statement when he told his hearers that they had given up their rights in the "great default," and when he called on them to assert their power through organization and group action. Everywhere throughout the areas of the study clubs the purchasing power of the consumer was hailed as the way to put the consumer in the driver's seat, as Dr. Coady had predicted. Sister Marie Michael, speaking at the women's conference in 1942, finished the analogy by describing the driver as a woman on her way to the co-op store.

Every purchase made at the co-op became identified as one of Dr. Coady's famous "little blows, destined to end up in the sledge hammer lick, powerful enough to command ownership of manufacturing by co-operators."

Exhortations at meetings were not always sufficient to convince women that they should be completely loyal to the co-operative way of buying. While we continued to do our best to promote the co-operatives we challenged directors and managers to remove the obstacles that stood in the way of greater participation by women in the direction of co-operatives.

The women believed that they should have some say in the appearance and layout of the store and on the service and the kind of merchandise offered for sale. Loyal co-operators saw many problems that constituted a hindrance to the attraction of new members and to the satisfaction of those who were persuaded to join.

It was easy for these women to give examples of what they called poor service. The manager of one store which was in difficulty because of a large inventory, could not understand why the bed sheets on the shelves were not selling. The women buyers could have told him if he had asked THEM why he could not sell the sheets, which were in the luxury grade, of the finest percale, hemstitched and embroidered. None of the women could afford them and so they went to the department store for unbleached sheeting which they made up on their sewing machines. One woman commented: "The old embroidery in my house is what I do myself and I don't embroider sheets!"

Another store, also with a high inventory, seemed to specialize in gadgets, most of them of no interest to shoppers. This manager outdid himself when he offered butter presses for sale. This interesting item was for the purpose of converting a lump of butter into a perfectly molded square pound. In a rural store this might have been of some use, but in a town? The women members had no cows, no churns and no use for a butter press.

There was no means of communication between the management and the members. At one time there was an effort to do this by means of a women's committee to be a liaison between the shoppers and the directors but the plan was a failure. The committee had no official standing, and a meeting with managers or directors appeared to be a special privilege conferred on it by a condescending group.

The women on the Extension staff were in a very good position to make recommendations that could solve these problems and help to ensure the loyalty of the co-operative shoppers. They urgently advocated that the co-ops, in joint effort with the wholesalers, should hire consumer consultants to plan and carry out in-store programs in co-operative buying and consumer education. We thought this was as important for the success of a co-op store as was the hiring of a general manager. We believed that such a project would soon pay for itself by helping to increase the volume of business.

The second recommendation was that efforts be made to clear the way for women to become directors of co-ops. As early as 1938 Sister Marie Michael, in a speech given at the Rural and Industrial Conference, had expressed the hope that "very soon" there would be a woman on the board of every co-op in the province. This did not happen.

In retrospect it may seem strange that, while so much importance was attached to the power of women in the co-operative movement, so little effort was made to ensure that they would have a place in directing the policies of the co-op stores. It was disappointing to find an all-male board directing a business that depended almost entirely on the support of women buyers. There were some women's names on the list of charter members of co-operative stores. There were none on the rosters of the first boards of directors. It was generally accepted that women could choose to determine the success or failure of a co-op store but it was as difficult to elect a woman to the board of directors as it was to send a women to the House of Commons.

It may be that one reason for the failure to get women represented on co-operative boards could be found in the social attitudes of the times. It was uncommon to find a woman in public office. Not too many years had passed since Canadian women had been granted the franchise. This had happened only after an august body of men in the Privy Council had made the momentous decision that a woman was a person. The election of a woman to anything more important than the Ladies' Aid was a novel idea.

It is interesting that one of the early credit unions had a little hesitation in permitting women to become members because they had no income of their own from which to save. Married women had no personal income because they did not engage in gainful employment. They could save only from their husbands' pay. It took consultation with the highest authorities in the credit union movement before women were allowed in that credit union, "with all its privileges."

The status problem extended also to the co-op store. Married women spent their husbands' pay either in co-ops or other stores. They had no money of their own with which to buy shares, but that seemed to us a flimsy excuse to keep them from becoming directors.

But there were stirrings of democracy. In the study clubs the principles of open membership and democracy in co-operation had been taken to apply to men and women alike. The principle was one member, one vote and not one man, one vote. How to apply the principle in practice so as to give the women membership in the society and the right to sit on the board caused no end of debate and discussion. The men, firmly entrenched on the boards, despite their efforts to persuade women to be loyal to the store, were not astute enough to see that the best way to ensure the success of the co-ops was to make it easy for women to become directors.

Instead of vigorous action to this end they wasted time pondering such weighty questions as: Who should hold membership in the co-op, the woman who spends the family income or the man who earns it? It made sense that the woman whose purchases kept the store in operation should be a member, have a vote at the meetings and be eligible to run for office. But if this were achieved, would the man of the house be disenfranchised?

The solution to the problem seemed very simple to the women. Husband and wife should both be members. It was always amazing to us that such a simple solution would seem impossible to the men. Our proposal gave rise to more questions. Could a family afford to buy shares for two members? And who would be credited with the purchases for the purpose of distributing patronage rebates? The rules governing eligibility to be a director required a specified amount of purchases during a given period in the fiscal year.

To overcome this difficulty the women proposed that a co-op amend its by-laws to allow the wife of a member to obtain membership by purchasing one share. This would give her voting privileges. If she desired to run for office, surely the question of assigning the required purchases to her could be settled amicably within the family circle.

Another proposal was for joint membership, but this caused much dissension. In the case of joint membership the person whose name appeared first on the register would have the voting privileges. What serious family disagreements were foreseen in the decision: "Who's on first?"

It seemed to us that little progress had been made since 1844 in the recognition of women's rights in the co-operative movement. One of the Rochdale Pioneers was a woman, and the Toad Lane store had a good percentage of women members.

Holyoake had reported that the Rochdale store rendered valuable aid in realizing the civil independence of women. In his amusing way he wrote: "Many married women became members because their husbands would not take the trouble. Others joined in self defence to prevent their husbands from spending their money on drink. The husband cannot withdraw the earnings from the store standing in the wife's name, unless she signs the order."

The smart women of Rochdale did not, however, have complete control of their store earnings, because the husband could get the money through a court order. But Holyoake had a solution for that too. "A process takes time," he said, "and the husband gets sober and thinks better of it before the law can be moved."

Holyoake continues: "Many single women have accumulated property in the store which becomes a certificate of conjugal worth. And young men, in search of a prudent companion, consider that to consult the books of the store would be the best means of directing their selection."

Although women's participation as voting members of the Rochdale store was somewhat involved with alcoholic husbands and prospective suitors, it seemed to us to be easier to obtain than co-operative membership in the Maritimes a hundred years later.

The high point of the program concerning the "Woman with the Basket" came at a women's conference held at Antigonish in 1942. The decision to hold this conference had been taken at an Extension staff meeting a few months before. There were 143 delegates from co-op stores, credit unions, women's guilds, youth groups and other organizations interested in co-operation. The purpose of the conference as stated at the staff meeting was "to waken among women's groups an active interest in adult education and co-operatives." The women speakers stressed the importance of gaining the support of women in the successful pursuit of co-operative goals and the urgent need for a program of consumer education. They repeated in the strongest terms their belief that there was a need for women field workers to promote an educational program in the stores and for immediate attention to the problem of removing obstacles in the way of electing women to co-operative boards.

In addition to the women speakers there were three men on the program, Dr. Coady, J.T. MacDonald, manager of the Cape Breton Co-operative Services, and Dr. Dan MacCormack of the Extension Department. All were concerned about women's power to change the world. In closing Dr. Coady said: "This conference is designed to give you the green light on a road that has no speed limit — the road to progress."

The women's guilds represented at the 1942 conference had been formed as sort of a substitute for the representation of women on the board of directors. The co-operative guilds had a long history. They had been started in England and had spread throughout Europe and become international in scope. Their chief purpose was to give women a part in co-operative education and to be a sort of "social arm" to each co-op. In Sydney Mines the British Canadian Co-operative which had been founded by a group of co-operators from England and Scotland early in the century had a co-operative guild.

As early as 1935 when the study club members became interested in the guilds the women on the Extension staff helped to organize them. The women who started the guilds were genuine co-operators and, for a time, they did carry an educational work for their stores. But they became discouraged when the guild rooms which they had prepared and furnished became repositories for sacks of flour, old account books, empty cartons and other overflow items from the store.

Although we worked with the guilds we were not entirely satisfied with their purpose. We were afraid that the guilds would become auxiliaries to the directors, attending to the work of organizing social activities and raising funds.

There were historical grounds for our apprehension. Very often in the past men had set up women's auxiliaries to their societies to raise money and serve tea. We had evidence that the co-operative directors had such little chores in mind for the women's guilds. We thought that women's participation in the co-operatives should be direct and not by way of a secondary or auxiliary organization. In this we were supported by Dr. Coady who thought that guilds made women second-class co-operators.

Moreover, we were doubtful of the motives of men who were so anxious to promote women's guilds. Our suspicion that these men considered guilds to be a substitute for direct participation was confirmed at the founding meeting of the Nova Scotia Co-operative Union in 1944. There was a serious confrontation between men and women co-operators at this convention which dissolved the Nova Scotia Co-operative Educational Council in favour of the new union.

The convention adopted a constitution and by-laws. One section of the by-laws read: "The members shall have authority to elect as associate members women's guilds, educational bodies or other organizations interested in the promotion of co-operative ideals and practices, but these associate members shall have no power to vote *on any question arising at any meeting* (italics mine) held by or under the auspices of the Union."

The exclusion of the women's guilds, so highly esteemed by the men, was definite and complete. The co-operative societies had been invited to send women to the convention. These women held a meeting at which they discussed the situation in which they found themselves with respect to participation in the direction of co-operatives and they drew up a recommendation for presentation to the convention. They requested that the directors of the Nova Scotia Co-operative Union provide at the annual meetings a separate session of women's guilds, educational committees and other groups of women operating under the authority of or in conjunction with organizations that were members of the Union. They asked that recommendations arising out of this special session be presented to the annual meeting for discussion and appropriate action in which case the women's section would be entitled to *four* voting delegates.

This modest request for four votes on one item of the agenda at an annual meeting was rejected by the convention. The women left that convention with bitter feelings, seriously doubting the sincerity of the men who proclaimed the importance of women to the co-operative movement. One delegate said: "Either the men do not believe that we have any right to participate in the direction of co-operatives or they are afraid of us. They must believe that four women could wreck the Nova Scotia Co-operative Union!"

But in spite of the stand taken by the founding convention of the Nova Scotia Co-operative Union, expressions of faith in the contribution that women could make to the co-operative movement continued to come from all quarters. The record of board and annual meetings of the Union are filled with resolutions to "take steps" to involve women directly in the movement.

There was one promising decision at the convention. The delegates recommended that the directors study the subject of changing the by-laws of the co-ops to permit the wife of a member to obtain membership by purchasing one share. This resolution was placed on the agenda of the first meeting of the Union's board of directors. The discussion showed how far the resolution was removed from practical application.

One of the directors, a prominent official in the co-operative movement, was in great fear of the proposed changes in the by-laws. He said that the only "brake" against the possibility of two people from the same household running for office was the rule that candidates had to have a certain amount of purchases in order to qualify. "Now we are talking of removing that brake," said he. He feared that if the by-laws were changed, a husband and wife might be elected to the same board. What dire consequences might occur from this revolutionary happening he did not explain, possibly because Dr. Coady, who was present at that meeting, promptly silenced him.

Said Dr. Coady: "What if these two were the best qualified to be directors, perhaps the two smartest people in the community? Would you trade the woman for an incompetent man?" I had been elected to the board of the Co-operative Union and was present at that meeting. But, by a commendable example of prudence and humility, for once I held my tongue on the topic.

Eventually the recommendation was circulated to all the co-operatives, but there were no practical results.

By 1950 the debate was still going on. In that year Maritime Co-operative Services Ltd. appointed a special committee to study the question. The committee consisted of three women prominent in the co-op movement, Kay Desjardins, Nadine Archibald, Secretary of the Nova Scotia Federation of Agriculture, and Jean Munro, a Department of Agriculture worker. The following is from the committee's report:

> "When the co-operative movement was given its birthright was it intended that it should be dominated by one political group, one religious sect or the male species of the human race . . . ?
>
> We believe that, up to the present time and in view of what information we can obtain, one of the serious drawbacks to progressive development has been the omission of full and equal participation by women. We must conclude that men have dominated all phases of the movement, and the women have been on the outside looking in with no invitation to enter. We believe this to be one of the major violations of the philosophy of the co-operative movement."

The committee recommended, among other things, that efforts be made to obtain legislation that would permit joint accounts, thus enabling a member's wife to have equal participation in the direction of co-ops.

In that same year the Nova Scotia Co-operative Union passed a resolution stating "that this convention go on record as recognizing the place of women in the co-operative movement and that a committee be appointed from this convention to assist in having women take their rightful place in all fields of co-operative endeavour.

Two years later at the annual convention A.B. MacDonald, who was then general secretary of the Co-operative Union of Canada, gave a list of causes for the slow growth of urban co-operatives. Number two on the list read: "Women have not been brought into the picture."

The 1953 convention of the Nova Scotia Co-operative Union took as its theme "The Place of Women in the Co-operative Movement." All the principal speakers were women and they pleaded their case as they had done repeatedly over the years. Once again a special women's committee was appointed to make recommendations. The committee's report with the familiar recommendations was received, praised and forgotten.

It would probably be unjust to accuse the men of deliberately discriminating against women in co-operatives and that women were kept from participation in the direction of co-ops by men who believed in the inferiority of the female mind. In our charity we blamed apathy and a slowness to accept changes in the position of women in society.

In spite of all the disappointments it was heartening to know that we could depend on the support of the founders of the movement. Dr. Coady, who was always on the lookout for "smart" people made no distinction between male and female brains. He was our greatest booster and he encouraged us in all our endeavours. If the "steps" had depended entirely on him they would have been taken quickly and soon, and they would have been giant steps. Dr. Tompkins, too, backed us up. Free from any kind of prejudice, he put the answer to the question of women's participation in the co-operative movement in one sharp sentence: "Adult education must be prepared to find truth where it finds it; it must also be prepared to accept talent where it finds it."

Chapter XIII
A PROGRAM FOR SOCIAL CHANGE

The ultimate objective of the St. F.X. Extension movement was to bring about social change. Its leaders wanted to attack the causes of poverty and its attendant evils and remove them forever. They had a vision of a new social order in which there would be economic as well as political democracy.

The credit unions and co-operative stores were to form the basis of the new order because through them the people would acquire ownership that would give them a vested interest in the economy and a powerful voice in national affairs. Dr. Coady had said of the small co-ops in *Master of Their Own Destiny:* "The homely rugged tasks we do today . . . are but the solid ground that holds our feet the while we build our wings."

In 1978, the fiftieth anniversary of the St. F.X. Extension Department, a conference, much like the Rural and Industrial Conference that marked the beginning of the Antigonish movement, was held at the university.

At one of the sessions of this conference a leaflet entitled "A Program for Social Change" was distributed to the delegates. It set forth a summary of the goals of the St. F.X. Extension Department and listed the various fields of action necessary to bring about reforms. This program had been prepared for a Rural and Industrial Conference and it had been widely used throughout the Maritimes at short courses and meetings. Many of the participants at the 1978 conference had never seen the "Program for Social Change," but to the oldtimers in the movement it brought back many memories.

Sister Irene Doyle recalled how she had carefully lettered the original chart on a four by six foot piece of oilcloth and attached it to a wooden roller so that it could be easily transported from meeting to meeting and province to province. I had a good reason for remembering that chart. It did not fit in the trunk of a car and I frequently had to ride in the back seat holding "The Program for Social Change" at the proper angle so that it would not put out the window. That chart travelled to all the short courses in the Maritimes. It was a nuisance to carry, but we could not have run the short courses without it.

In 1978 the Program for Social Change seemed as relevant as in 1940.

The fields of action listed in the "Program" were: labour unions, finance, production, consumption, marketing, services and cultural activities. The chart forecast that the credit union movement would grow from the provincial, national, and international associations to a national co-operative credit society and co-operative mortgage banks. Co-operative marketing and consumer co-operatives would lead to the establishment of co-operative factories. The vocational groups were represented in the program in their relation to one another, the primary producers and the industrial workers in their own special organizations, and all working together in consumer co-operation.

The Extension Department at that time was directing special programs to the labour unionists. In the industrial towns, during the fall and winter study sessions, it held evening classes on topics of special concern to organized labour. It arranged radio lectures by teachers from the university. The Extension libraries had good selections of books on labour topics and the Maritime Co-operator's labour page served to unify the program directed to labour.

The wage earners had been quick to form credit unions but had not shown the same energy in building and supporting co-operative stores. At every opportunity the Extension workers used the theme: "The workingman who stops at negotiating a higher wage and makes no attempt to control prices through consumer co-operation is fighting with one hand tied behind his back." We used to build air castles about the good that would come from the use of "labour's other arm."

Progress had been good in three of the services listed in the program — insurance, the press and housing. But the intrepid co-operators had failed in two ventures — co-operative medicine and co-operative burial services.

Twenty years before the coming of medicare to Nova Scotia the study clubs were discussing co-operative medicine. The subject had been studied at a Rural and Industrial Conference. In an address on the topic Peter Nearing had said: "We are still in the horse and buggy stage of medicine while the times demand streamlined delivery."

At that time there was no talk of socialized medicine. The people who dreamed about co-operative medicine had a different concept. They thought that a group of co-operators could organize a medical clinic owned by the consumers who would prepay the costs of medical care on the insurance principle that everybody would not become ill at the same time. The co-operative would own the clinic and equipment and pay the doctors' salaries. There were models to imitate. There were successul co-operative medical clinics in the United States and physicians from these clinics had agreed to help the Nova Scotia co-operators get started.

As was expected, interest in the plan was highest in the industrial area of Cape Breton, where for many years workers had prepaid medical care through a weekly check-off from wages. In the mining industry the check-off for doctors was a condition of employment. The plan was a one-sided contract and the consumers did not have any control of the service. As a preliminary to the discussions on co-op medicine, a servey of the medical check-off systems was made by Kay Thompson and published under the title: "Medical Services in Glace Bay."

Because the Cape Breton industrial workers were accustomed to prepay the costs of medical care a change to co-operative medicine seemed logical to them. The co-operators saw no difference between controlling money through credit unions, merchandising through their co-operative stores and controlling their medical services through a co-operative clinic. They were especially interested in the American co-operative medicine plans because the clinics made group medicine and specialization possible.

Although these co-operators did not succeed in their purpose, they must be admired for their bravery. The first hurdle was to get an amendment to the legislation under which the province's physicians practiced. The start was made in New Waterford. The New Waterford Co-operative Health Services Association was organized. It became incorporated and elected a board of directors. Dr. Shadid, one of the pioneers in co-operative medicine in the United States, agreed to come to Nova Scotia to meet with the physicians, the hospital boards and the Minister of Health.

The plan failed because the doctors refused to discuss anything with the Association. They called in a representative of the Nova Scotia Medical Society who was hostile to any plan that would disturb the status quo. The Association, whose members had spent countless hours in study and planning, had to be dissolved.

The effort to organize a co-operative medical clinic was doomed from the start. Control of the distribution of medical services by consumers was impossible because the vested interests were too strong. The opposition was insurmountable. But the attempt is interesting because it demonstrates to what extent the early co-operators were absorbed in exploring the possibilities of group action.

The leaders in the Antigonish movement believed that they would see people change. Study and group action would lead not only to ownership in co-operative institutions but also to the development of confident, self-reliant citizens dedicated to the work of changing society and competent to do so. Participation in co-operative business would prepare the people for intelligent political activity which was another form of group action. Proud owners of their homes, their banks, their retail and wholesale business and their factories would be well equipped to determine for themselves the kind of society in which they lived.

It may sometimes have been difficult for observers to take note of the big plan because there was so much busy-work in forming and running the co-operatives. Attention to the beginnings may sometimes have obscured the vision of the future. But the grand plan was there from the start. Dr. Coady said: "Through credit unions, co-operative stores, lobster factories and sawmills, we are laying a foundation for an appreciation of Shakespeare and grand opera."

Chapter XIV
THE KEYS OF THE CITY

The Antigonish experiment was something new and exciting in adult education. It may not have been the first time that people had formed clubs for the purpose of studying some subject. The St. F.X. Extension program was different because of the way in which it involved large numbers of people, all studying the same topics, under a general plan, and all looking forward to definite co-operative action.

It was not long before news went out of Antigonish. The leaders of the movement may have been somewhat surprised by the attention which the program received from educators in all parts of the American continent.

Prior to 1933 the movement attracted attention mainly in the Atlantic area and in some of the western provinces. But, beginning in 1934, there was a burst of publicity which brought inquiries and visitors from Central Canada, the United States and abroad. An article in *C.K.'s Weekly*, London, was the forerunner of many other articles. Feature stories appeared in the *New York Times*, the *London Times, America, The Commonweal, Le Devoir,* the *Montreal Standard* and many farm journals. Some publicity, like an article in *Coronet*, was fulsome and inaccurate to the point of embarrassment.

Letters began to pour in from all quarters and the letters were followed by visitors who wanted to see for themselves how the program was carried out. The leaders of the co-operative movement in the U.S. were particularly interested and many of them came to the annual rally of co-operators at the Rural and Industrial Conference.

The list of distinguished visitors to the Rural and Industrial Conference contains the names of the most prominent U.S. co-operative leaders. Among them were: E.R. Bowen, secretary of the Co-operative League of the United States; Murray D. Lincoln, secretary of the Ohio Farm Bureau Federation; Charles A. Beard, of the American Association for Adult Education; Elmer Scott, Executive Director of the Civic Federation of Dallas; Reverend James A. Byrnes, Executive Secretary of the U.S. Rural Life Conference; Michael Williams, editor of the *Commonweal*; Reverend L.S. Ligutti, Dorothy Day; Dr. Benson Y. Landis. These are only a few of the educators who found their way to Antigonish.

Some of these returned with large groups for the purpose of making a tour of the co-operatives and credit unions. Dr. J. Henry Carpenter of New York led several groups, one consisting of clergymen. In another group seventy-two Quebec farm leaders made the rounds of the Nova Scotia co-operatives.

The first and most important feature of a visitor's stay in Antigonish was a conference with Dr. Coady who was always happy to give a full explanation of the history and philosophy of the Extension movement, and who put as much energy and enthusiasm in the conversation with one or two persons as though he were addressing an auditorium full of people.

The visitors then conferred with A.B. MacDonald, who knew every detail about the study clubs, the credit unions and the co-operative stores. Kay Thompson, who looked after the innumerable activities connected with running the hundreds of study clubs, answered the questions about the clubs and the varied amount of study material that was supplied to them. Sister Anselm kept the Extension show piece, the map of the Maritimes, dotted with coloured pins which showed at a glance where a credit union or a co-op store or a lobster plant was located. As time went on Sister Marie Michael gradually assumed the duties connected with the reception of visitors.

Those visitors who were meeting Dr. Coady for the first time were usually overwhelmed by the shock of "intellectual dynamite." Some visitors came because they had heard Dr. Coady speak and wanted to learn more about the practical application of his ideas. One young couple bicycled all the way from New York after attending a meeting at which Dr. Coady had been the guest speaker. Pierre Eliott Trudeau, on his bicycle, was one of the early visitors.

For some of the early visitors the days spent at St. F.X. were sufficient and they went home satisfied. Others ventured further away to the industrial area of Cape Breton where they expected to meet the people in the study clubs. The Extension office in Glace Bay was a good place for this purpose. On any day the visitor could be sure to meet credit union members or officers and men active in the labour unions.

The industrial area was interesting to visitors who knew something about its history and who were looking for information on the effect of the Extension movement on communities that had been hard hit by the depression. Sometimes we got the impression that a visitor expected to find wild-eyed revolutionaries lurking about the Extension office. They had come with previous knowledge of strikes and May Day parades. It made us happy to watch these visitors meeting well-spoken men and women with whom they had serious, intelligent and satisfying conversations on social theory.

Finally, in Cape Breton there was Dr. Tompkins. No survey of the movement was complete without a conference with this pioneer adult educator.

Our visitors welcomed the opportunity to attend a meeting of a study club or of the Associated Study Clubs. Since a meeting took place almost every evening somewhere in the area even a short visit enabled a visitor to take part in one. Occasionally, we called a special meeting for a visitor who had something important to communicate.

Although the visitors enjoyed the meetings they sometimes found the journey to a meeting place somewhat trying. We once brought a professor from an American university to a meeting in Main-a-Dieu. It was late autumn and the roads were rough and muddy. It was also a very dark road as the only illumination when the city of Sydney was left behind came from the moon. When the moon was full the going was great, but in periods of lunar decline the travellers went along in the dark, assisted only by the car headlights which were not very powerful in the early thirties. On this particular evening the moon was off duty and the combination of the black night, dangerous curves, rough roads and a driver given to speeding rendered our guest speechless with fright.

On the return trip conditions were not any better, but the experience of the meeting, which he thoroughly enjoyed, restored the faculty of speech to the learned doctor. In defence of timid visitors it must be said that some of the drivers would never have won a prize for safe driving. It is a wonder that we escaped injury on the many excursions into the countryside. Perhaps we survived because there was so little traffic. Cars were scarce on country roads at night. The car owners were safe at home asleep.

Advance information about an impending visitor was received with interest especially when we expected to receive persons who were connected with co-operatives and whom we could put on the program for a meeting. Once we were advised that an English lord was on his way. As we had never before chatted with a nobleman, we had to look up the etiquette book to learn how to address him. But the men from Phalen local of the U.M.W. were not impressed. His Lordship had requested permission to visit a mine and the men who were to accompany him gave him no special treatment. "He'll come up to the surface with a dirty face, like the rest of us," was the only comment.

During the thirties we had interesting experiences with the large co-operative tours. The Co-operative league of the United States organized an annual tour of approximately one hundred and fifty persons who began by attending the Rural and Industrial Conference. This was followed by a few days of special conferences at Antigonish and then the tourists went on the road, visiting the credit unions and co-operative stores in Antigonish and Guysborough counties and Cape Breton Island.

The planning for the tours was done at the Extension office in Antigonish, but it fell to the field workers to form an advance party to prepare the way and to act as trouble shooters during the tour. This preliminary work would probably be called a P.R. job today, but in the thirties we knew nothing about public relations. Those were pre-causeway times, and it is doubtful that knowing we were in public relations work would have made it any easier to get that large number of cars across the Strait of Canso on time, while anxious cooks were keeping dinner hot on the Cape Breton side.

One purpose of the advance trip, made some time before the tour, was to impress upon directors of credit unions and co-ops the importance of polishing up the premises so as to make a good impression on the distinguished visitors. In one case we succeeded beyond our expectations. The directors of the co-op store decided to paint the store in honor of the visitors, but they could not afford to paint the four sides. The directors held a meeting and devised a plan by which they would use paint on the front of the building and whitewash on the other three sides. The usual way to use whitewash was to mix it in a bucket and apply it with a wide brush. The ingenious directors mixed the whitewash in the accustomed way, but they then poured it into an empty paint can left over from the front job, being careful to leave the brand name prominently displayed. They then proceeded to apply the whitewash with a paint brush, thereby giving the impression to all who passed by that they were painting and not whitewashing. The reason for this deception was that the community's general store was directly opposite the co-op, and the wily directors did not want their competitor to know that they were too poor to paint the co-op store all the way around.

On the whole, the tours were successful and the problems along the route were fairly easily overcome. We could do nothing about the dusty roads and there were complaints, especially from those tourists who suffered from allergies to Nova Scotia dust, and there was a good deal of sneezing and coughing. But then a kind providence would send a gentle rain to lay the dust and raise the spirits of the co-operative pilgrims.

Finding overnight accommodations was the most difficult part of the arrangements. One year a travel schedule had to be completely changed so that the overnight stay was shifted to Cheticamp from the location which had been previously agreed upon and where arrangements had been made for overnight. A.S. MacIntyre and I were dispatched to Cheticamp to find accommodations for the tourists who were coming in the opposite direction over the Cabot Trail. We arrived in Cheticamp late Sunday morning with only the afternoon in which to find beds for over a hundred people. It was impossible to find enough room in the inns which were already filled with guests since it was the height of the tourist season. We spent the afternoon knocking on doors and, although there was a very short time to prepare for overnight guests, we were cordially received. Our job had been made easier by the parish priest who had announced from the puplit that the onslaught of the co-operators was imminent and had asked the parishoners to receive the strangers graciously.

The tourists greatly enjoyed the meals that were served enroute, especially the fresh fish easily available in August. The women of the community where the caravan stopped would serve a delicious meal in the community hall. These were friendly gatherings with speeches of welcome and short talks by local leaders. The guests would contribute to the program with information on the co-operatives with which they were connected. In one of the welcoming speeches the parish priest grew more and more eloquent as he went on, ending in a burst of enthusiasm in which he offered to the guests the keys of the city. But he followed this grand gesture of friendship with the rueful announcement: "But we have no keys." Some of us thought this *faux pas* was very funny, but the tourists loved it and gave him wild applause. Even without keys they knew they were welcome.

The tours were held in August. One year it was a particularly warm month and the temperature in Antigonish was in the high eighties. Some of the women had arrived from the United States equipped with heavy sweaters and skirts, woollen stockings and blankets. They were well prepared to withstand the icy Canadian winds sweeping down from the Arctic. It had not occurred to them that the inhabitants of this frigid country would have blankets to spread over their shivering guests from warmer climes. Some good came out of the miscalculation about temperatures in Nova Scotia. The

Antigonish shops sold out their entire stock of warm weather dresses.

In the Glace Bay office we sometimes had a visitor on whom we would not have bestowed the keys of the city. This might be a reporter whose only motive in taking up our time was to get paid for a story. We developed a kind of sixth sense that enabled us to spot a phony. That was the signal to become taciturn and divulge as little information as possible. We usually found out later that our reticence mattered not at all and did not hamper the scribe's effort.

We once had a reporter who wrote about the heavenly blue eyes of one of the fishery field workers. We had never noticed the colour of his eyes and we eagerly awaited his return from a trip to find out if his eyes were blue, and sure enough they were. That reporter had one item correct.

The writers looking for a good story exaggerated what they observed. For example, the story in *Coronet* emphasized the poverty of the region, especially Antigonish. It said of the Antigonish people: "Their faces are pinched and chalky. *They never smile."* (Italics mine.)

They also exaggerated the results of the movement and this caused some embarrassment. Nobody knew better than we that the whole economic structure of Nova Scotia had not been rebuilt.

But the phonies were the exception. The visitors we took to meetings were interested and interesting. It was a pleasure to meet them and we remembered them long after they left us and we had caught up on the work neglected while we talked with them. At Antigonish the reception of visitors became part of the daily duties of the staff. The genial Sister Marie Michael became accustomed to greeting visitors from all over the world. She kept a guest book of these visitors with little notes about the circumstances of their visits.

The visitor who made the most impact on the movement was the one who came to call and stayed. If the keys of the city had been conferred on any one, it should have been on Mary Arnold, a director and the treasurer of the Co-operative League of the United States. She came to the Rural and Industrial Conference and became enthusiastic about the Extension program. Shortly after, she returned to Nova

Scotia to work for the St. F.X. Extension Department in co-operative housing, a field in which she was an expert.

Mary Arnold was a true pioneer who opened an entirely new field of co-operative endeavour in Nova Scotia. The task she set for herself was tremendous. The Toad Lane Study Club of Reserve consisted of men who were inspired and directed by Mary Arnold to build Tompkinsville, the first co-operative housing community in the province. They had little formal education, no building skills and no money. Their assets were their experience in group action in the study clubs and credit union, an understanding of co-operative principles, an overwhelming ambition to be the owners of their homes and the will and energy to do hard physical labour. Tompkinsville was built in 1938, one of the worst years of the depression. How poor these co-operators were can be shown by the fact that, on one occasion, nobody in the group had twenty-five cents to pay a taxi for Joe Laben to go to New Aberdeen where he was expected at a meeting of a new housing group. Eleven families with a total of forty-eight children, and not a twenty-five cent piece among them!

Mary Arnold taught them how to draw plans and blueprints. Under her direction they built scale models of the houses they hoped to obtain. She led them through the complicated processes of acquiring land, organizing the housing association, becoming incorporated, and learning how to run the business of the association. She was the liaison between the group and the Housing Commission.

When building time came she was the expert who attended to the buying of material. It was often said that Mary Arnold had in her head a count of every stick of lumber and every nail that went into the Tompkinsville houses. And she was there on the construction site from the first day when the basesments were dug until every house was ready to receive its jubilant occupants.

Mary Arnold and her companion, Mabel Read, came from New York City and settled in Reserve with complete acceptance of the Cape Breton way of life. Their home was a centre of education where the men worked on the technical and business details of the housing project. The women, too, held their meetings at the Arnold house. They made curtains and quilts and pored over house plans. Mary Arnold was receptive to all the ideas. She was surprised when Mary Laben insisted on a pantry in her new house. The pantry was a room which had gone out of fashion, but Miss Arnold said that if Mary wanted a pantry, then she should have it, fashion or not.

The first building project was a test house set up to provide information on costs and other aspects of the project. When the test house was finished Miss Arnold moved in and so did the study clubs on meeting nights.

It was a happy coincidence that Mary Arnold's experiment in co-operative housing should be in Reserve where Dr. Tompkins was busily promoting adult education and regional libraries. The two had much in common. Mary Arnold entered with enthusiasm into the general educational movement in the community. She was present at all the Associated Study Club meetings. One of her special interests was the promotion of health through good nutrition and she was very helpful in our nutrition program.

Although Mary Arnold appeared very serious and even austere, she had a good sense of humour. She derived much pleasure in writing and producing a play, "The Miner's Wife" which depicted the life of the miner's family. It provided a lot of fun for those who took part in it and it was also an excellent piece of co-operative education. It enjoyed successful presentations throughout Cape Breton.

She was very efficient and liked to have things done on time. She had no patience with the law's delay and sometimes communication with the officials in Halifax tried Mary Arnold's soul. She and the provincial registrar of joint stock companies had their difficult moments. But on the triumphant day of the opening of Tompkinsville, all was forgiven in the joy of the realization of a dream. Mr. Beasley crowned the speeches with a tribute to Mary Arnold. Those of us who knew of the differences between those two enjoyed how he paraphrased a well-known quotation. He said: "Although Miss Arnold and I did not always have two minds with but a single thought, we did have two hearts that beat as one."

How fortunate co-operators in this province were that Mary Arnold who came to visit stayed a while.

Chapter XV
WAITING FOR THE SNOWPLOW

Memory may be somewhat dimmed about some events of the years in field work, but not about what it was like to travel throughout the Maritimes in the thirties and forties. When Dr. Coady recited a list of the "good things" which rural people should enjoy he always included paved roads. How fervently we echoed his wish!

In spring and fall the dusty roads of summer turned into mud which made travelling difficult. In spring when the frost came out of the ground the bogs made some roads almost impassable. Sometimes a horse would be found nearby to pull the co-operators' car out of the mud. If that noble animal could not be found, the passenger who was fleetest of foot would be dispatched to the nearest telephone to summon help.

Because of muddy roads I missed the best advertised meeting in the history of the study clubs. The meeting was to be held at Gabarus, a rural community about twenty miles from Sydney. A week before the meeting the whole countryside turned out for the funeral of a beloved minister. The energetic organizer had enlisted the help of the local school-teacher who had her pupils write individual notices concerning the meeting. These were distributed to every person in the congregation.

The date was December 8 and winter had set in early that year. Freezing and thawing had made the road so bad that the driver decided to make a detour. We were looking for a lighted lantern which was to be placed on a schoolhouse gate post to guide us to the right spot. But the detour made the driver lose the way and, after wandering in mysterious paths for two hours, we arrived at the schoolhouse at ten o'clock. The people had gone home, taking their lantern with them.

I was particularly disappointed at my failure to get to this meeting because the planning had brought about my only personal encounter with male chauvinism during my years with the Extension Department. The arrangements had been made with A.S. MacIntyre to take this meeting. But when MacIntyre had to go out of town on that date I agreed to substitute for him. We often did this when one or the other had a change in plans. When the organizer came to the office to check on the arrangements he was horrified that I was to go to the meeting instead of A.S. When the initial shock wore off he grudgingly said: "I suppose it will be all right if she brings the speaker with her." I regretted that I did not have the chance to show him that I was really the speaker.

In winter it was snow that made travelling a nightmare. It blocked the roads and isolated us for sometimes as long as a week. That was when we practised the penitential exercise known as waiting for the snowplow. These plows were remarkably inefficient and unreliable. When the roads were piled high with snow it was not very comforting to learn that the plow, which should be out there clearing a path for us, was sitting in its garage, waiting for a part to come from Toronto where it had been sent for repairs.

Once we had a week of waiting at Canso for a plow to come to our rescue. A short course which started on a Monday morning came to an abrupt end on Wednesday when a record-breaking snowstorm blew up. The instructors were snowbound for a week. Dr. Coady was imprisoned in the rectory with the parish priest. A.B. MacDonald, two representatives from the Department of Agriculture and I were at the local inn. The telephone was our only link with the outside world and, after a fourth day, that was cut off.

A.B. MacDonald had brought a movie for the local Red Cross to use as a fund-raising project but the event had to be cancelled because of the storm. To help pass the tedious days, A.B. showed that movie every afternoon in the inn parlour. Fortunately it was a musical picture and we could enjoy the repetition of the music of the Philadelphia Symphony Orchestra. The Red Cross ladies had made large quantities of fudge which they were to sell at their movie evening and the inn had a good supply. We ate huge helpings of fudge as we listened to Leopold Stokowski and Deanna Durbin while the storm raged outside.

On the seventh day of our penance, as we were sitting down to dinner, we heard the unmistakable sound of the snowplow. We left the table, dinner untouched, and fled.

While the experience of being snowbound indoors was unpleasant, it was not as bad as being outside in the midst of a storm. Once we had to leave the car and stand outside for some time during which the weather changed from snow to rain and then to sleet. My stockings froze to my legs. I expected to develop pneumonia from that ordeal, but I did not get a sniffle.

One of my worst experiences with snowstorms was on a journey from North Rustico, Prince Edward Island. About mid-week during a short course a severe snowstorm halted everything and continued for the rest of the week. After much deliberation and consultation with the weather office it was decided that we would go by sleigh to Hunter River, there to board a train for Charlottetown.

I was staying at the local convent. On the day of departure I arose with the nuns at five o'clock in the morning. The Sisters were worried about my survival on the trip and they put forth every effort to see that I made it to Hunter River without frostbite. First I had a substantial breakfast with several cups of "good strong tea." Everybody was eager to contribute some article of clothing and I ended up wrapped in shawls, scarves and mittens. With the promise of prayers for a safe journey the Sisters saw me off in the black night and the shrieking wind.

We made our way slowly through the fields because the road was blocked with snow. The men walked over the worst parts of the way. This was before the era of "women's liberation" and I was forbidden to leave my place in the sleigh. Today, under similar circumstances, I would have the liberty to walk all the way from North Rustico to Hunter River.

A.B. MacDonald had forgotten his overshoes at Mount Stewart where we had held a short course the previous week. He had to trudge through the snowbanks with no protection for his feet which soon became numb with cold. He entered one of the farmhouses on the way and persuaded its residents to give him a long pair of woollen stockings which he put on over his shoes and fastened to his trousers with safety pins. He told us that he had taken the pins from the baby's diaper because they were the last two pins in the house. We did not quite believe that. We thought it was an embellishment of his yarn. The story of that journey has been told with many embellishments in words and in print. This is the first time it is told truthfully.

The journey to Hunter River, a distance of eight miles, took five hours. There we boarded a train for Charlottetown. This part of the journey was uneventful and, in our dilapadated state of dress, we had lunch at the Charlottetown Hotel. News of our arrival had preceded us and we were met there by Dr. Murphy, president of St. Dunstan's University, and Dr. A.E. Corbett, president of the Canadian Association for Adult Education and one of the chroniclers of the famous journey.

For some strange reason which I have forgotten we then found that we had to return to Hunter River from where we had just come. Since ordinary cars could not pass through the snow, the bookmobile was commandeered. I rode in the cab with the driver. In the van were four very large men, A.B., Dr. Murphy, Dr. Corbett, and Dr. Croteau, all armed with shovels. Thus escorted, we reached Hunter River in the late afternoon to find that the only transportation to Borden was a freight train. We did part of that journey on a flat-car which (to enlighten those who have not travelled in this way) is completely open to the weather since it has neither sides nor roof.

At Hunter River a silent stranger attached himself to our party. He was a tall man dressed in black with a black homburg which remained secure on his head during the rest of that wild journey. We learned that he was a tea salesman, but after divulging this information he spoke very little. But he was a lifesaver for me. Wedged between him and A.B., standing on that open flat car, I was preserved from being hurled into the Prince Edward Island potato fields.

At Borden A.B. ventured forth to look for a conveyance that would bring us to the ferry. He found a man with a truck who was daring enough to attempt it. We crossed over to Tormentine where we expected to find the car which had been shipped from Rustico at the first snowflake. The car was there, but it could not be removed from the turntable and put on the road until the arrival of the morning shift. We therefore spent the night in the railroad station with the night clerk for company. The tea merchant's hat never left his head.

The ride to Antigonish was smooth and there I got on a bus bound for Sydney, arriving at midnight. The journey by sleigh, train, bookmobile, freight, flat-car, truck, ferry, automobile and bus had taken over forty hours. I vowed to have nothing more to do with the Extension Department or the co-operative movement. Two weeks later I was in New Brunswick battling another snowstorm.

I missed one unusual encounter with snowbanks. I was to go to a meeting in Main-a-Dieu with Marjorie McKinnon, a home economist on the Extension staff, but I became occupied elsewhere and Marjorie went alone. We were to stay overnight and return to Sydney the next day. During the night a fierce storm blew up and the road was completely blocked. Next day the road was opened by manpower, but one huge snowbank defeated the shovellers. By the afternoon a plan had been devised to get Marjorie home. She was driven to the foot of the snow mountain by horse and sleigh. Suitable tests were made to confirm that the snowbank had hardened enough to bear her weight. She climbed to the top and the rest was easy. She slid down the other side. Once started, nothing could have stopped her. At the foot of the snowbank another team was waiting to take her home.

Once when I was very anxious to get home quickly from a short course in Arichat, Angus Rankin, the agricultural representative, found a way for me to get to Sydney by car. I was not at all perturbed when I discovered that my ride was to be in a bread truck. Anyone who had travelled on a flat-car in a snowstorm was not likely to shrink from riding in a nice warm bread truck. I was somewhat displeased, however, when before heading for Sydney, we had to make the rounds of all the stores on Isle Madame, delivering bread. Angus had made an error. The truck was coming from, not going to Sydney when Angus had made the arrangements with the driver. We reached Sydney long after the bus and the train and after all street-cars to Glace Bay had been put in the car barns for the night. It was sub-zero weather and the truck had no heater. It was some time before I found it in my heart to forgive Angus.

In early spring when the roads were at their worst the bus company would, for a short period, substitute a seven passenger car for the regular bus. I was once the seventh rider in a car that carried five inebriated sailors on leave. It was a harrowing trip. The sailors quarrelled and sang by turns, and I heard words that I had never heard before and that I never heard again.

One exception to cold journeys took place on an ill-fated expedition to Scatarie Island. Rev. A.P. Poirier, parish priest at Main-a-Dieu, had arranged a meeting at which A.S. MacIntyre and I were to attend. The three of us, accompanied by Mrs. Piper MacMillan of Reserve, set forth by motor boat on a hot August day. After we had landed and Father Poirier had dismissed the operator of the boat, with instructions to return for us in three hours, we found to our dismay that we had landed at the wrong end of the island. The meeting was at the other end. Instead of staying where we were, awaiting rescue, we foolishly decided to cross the island on foot. When we limped into the meeting place, after our long trek through rocks and brambles, we found nobody there. All had gone home.

In pre-causeway days, ferries were used at the Strait of Canso for cars and trains. Most of the short course took place during the war years and that made riding the night train a time of terror for a coward like me. Rumours of enemy mines in the Strait of Canso were numerous and frightening. When the train approached Point Tupper, I would rise from my berth and dress, putting on my coat and overshoes, and wait, in prayer, throughout the long process of engine changing and sailing across the Strait on the ferry. My friends at the Extension office were greatly amused by my preparations to meet an enemy mine. They would inform me that, if the *Scotia* were blown up, I would not be saved from drowning by my coat and overshoes, since I was unable to swim, with or without overshoes.

Travelling in bad weather taught me how to dress for survival. It is a pity that the pant suit for women had not come into fashion. Dresses and coats were knee length, leaving a section of leg to be frostbitten.

There were two articles of clothing in current fashion that helped to preserve me from freezing. One was the velveteen overshoe, lavishly trimmed with rabbit fur around the top, and fitted with laces that could be tightened to make a snug fit. The other was the turban, a creation of wool jersey in a long strip that wound around the head and covered the ears, secure against the strongest gale. The only way that turban could get away from you was to have your head go too. I added my own favourites to these fashionable articles of clothing. These were mittens knitted with double yarn, and a blanket to keep away drafts in buses. Thus attired in all these accessories and my fur coat I was ready to brave the cold winds which Dr. Coady said made Maritimers tough and fit to meet adversity. In this I disagreed with him. The cold winds did not make me tough. They just made me shiver in my overshoes and mittens and wish that I had been born in Hawaii.

Chapter XVI
"A TERRIBLE BELIEF"

The St. F.X. Extension movement had certain built-in factors for success — its fundamental philosophy of brotherhood and mutual help, the genius of its founders and the dedication of its staff workers. But without the contribution of hundreds of volunteers the Extension program, in both its educational and economic aspects, would not have progressed beyond the first stages.

Among these volunteers were local clergymen who promoted and encouraged study clubs and the formation of credit unions and co-operatives. In rural and fishing villages these reformers directed the work of organizing marketing agencies and fish plants. Dr. Coady referred to them in *Masters of Their Own Destiny*; he noted that a few of them were engaged in promoting co-operatives long before the Extension Department was started.

The staff members were also cheerful volunteers. They paid little attention to the length of the working day. They were available when work had to be done in evenings and on weekends. They took holidays when there was a lull in activities. It was not unusual for a field worker to spend a full day in the office, dash home for a quick evening meal, leave for a meeting, and return as late as eleven o'clock.

The agricultural representatives were also valuable volunteers. It is true that the Extension program was admirably suited to their jobs in the farming districts. Credit unions and co-operative marketing and co-operative stores complemented their teaching of scientific agriculture. But these government employees put in many long hours of overtime and never counted them. They were present at many conferences where evaluation of the program and planning for activities took place and their contribution was always of much importance.

The heroes of the movement were the directors and managers of the credit unions and co-operatives. The credit unions, in the first years, were run entirely by volunteers.

Managers who were responsible for conducting the credit union business gave their services free. This is true of all the credit unions organized in the thirties. For example, Bergengren credit union at the end of its first year in business made a "gift" of five dollars to the manager who had presided at all the business hours and taken care of the accounts. The gift amounted to a remuneration of ten cents per week. The hourly rate is too small to merit reporting. At the end of the second year this credit union was able to increase the manager's gift to ten dollars.

The manager of Coady Credit Union received no remuneration for the first three years. The secretary who remained in that office for twenty years and who was the manager's assistant from the start received his first pay, fifteen dollars a month, in the credit union's sixth year. With the manager he spent every Sunday afternoon working on the credit union books. These two were also the janitors. They took turns at lighting the coal fire on business days in the unfinished building where the ink was frozen in the bottles and the blotting paper firmly attached by frost to their work table.

Glace Bay Central Credit Union's first payment to the clerk who assisted the manager was two dollars per week with the provision that any shortage in funds should be deducted from this salary.

Conventions were called very early in the movement. Delegates often attended conventions at their own expense or at best for bare out-of-pocket expenses. In 1934 Glace Bay

Central Credit Union's directors advanced five dollars to each of two delegates to a League convention at Port Hawkesbury. They were instructed to present bills on their return and refund any money left over. The delegates did very well. They each returned two dollars from the funds advanced.

In 1940 Bergengren Credit Union sent two delegates to the Credit Union League annual meeting in Halifax for a total cost of $25.00.

Delegates shared rides and accommodation. When they had friends or relatives in convention towns they got free room and board. It was good that Nova Scotians, and Cape Bretoners especially, were blessed with plenty of cousins living in various parts of the province and outside.

In the late thirties Percy Pellerin, field worker in the fisheries program, went to Boston to investigate marketing possibilities for the United Maritime Fishermen. He walked long distances each day to stay with a cousin and save hotel expenses. He saved a total of eight dollars by returning to Nova Scotia by bus instead of by the more convenient and comfortable train.

Directors of credit unions and co-operatives assumed a heavy responsibility in the conduct of business that was entirely new to them. It was a remarkable feature of the Extension movement that the members of the credit unions and co-operatives should unhesitatingly put their trust in these inexperienced bankers and merchants. As the business enterprises grew, so did the numbers of meetings to attend and the complexity of the problems to be met and solved. It must have often been tempting to stay at home on a winter night instead of taking a long ride to a meeting held in a cold hall.

Volunteers not only kept the study clubs going, not only manned the credit union wickets, kept the books, and served as directors and on countless committees. They also worked with hammer and saw, erected buildings and painted them. The labour they could provide free, but building materials cost money. When it became necessary the amateur carpenters laid down their tools to take charge of money-raising activities to pay for lumber.

Then there were the soldiers in the ranks, those who did not aspire to become directors, but who were the mainstay of the study clubs. Every credit union and every co-operative store had its boosters, who spread the gospel of the co-operative movement wherever they went. They formed a corps of itinerant teachers who never missed an opportunity to recruit a new member for a study club, a credit union or a co-operative store.

Dr. Coady was often heard to say that the people had to believe with a "terrible" belief in the ultimate triumph of adult education. The volunteers in the Extension movement must have been sustained by something approaching that belief as they worked so unselfishly to bring about a new and better social order.

Although the vital contribution of the volunteers may have been to some extent overlooked by chroniclers of the movement, Dr. Coady did not forget them. He dedicated *Masters of Their Own Destiny*, which is read in the far corners of the world, to the "unnamed noble souls who without remuneration are working overtime in the cause of humanity."

It was a privilege to know many of them, to work with them and to share their dream of a better world.